Contents

CYBERNETICS & HUMAN KNOWING
A Journal of Second-Order Cybernetics, Autopoiesis & Cyber-Semiotics

Cybernetics and Human Knowing is a quarterly international multi- and transdisciplinary journal focusing on second-order cybernetics and cybersemiotic approaches.

The journal is devoted to the new understandings of the self-organizing processes of information in human knowing that have arisen through the cybernetics of cybernetics, or second order cybernetics its relation and relevance to other interdisciplinary approaches such as C.S. Peirce's semiotics. This new development within the area of knowledge-directed processes is a non-disciplinary approach. Through the concept of self-reference it explores: cognition, communication and languaging in all of its manifestations; our understanding of organization and information in human, artificial and natural systems; and our understanding of understanding within the natural and social sciences, humanities, information and library science, and in social practices like design, education, organization, teaching, therapy, art, management and politics.

Because of the interdisciplinary character articles are written in such a way that people from other domains can understand them. Articles from practitioners will be accepted in a special section. All articles are peer-reviewed.

Subscription Information

For subscription send a check in $US (drawn on US bank) or £UK (drawn on UK bank or Eurocheque), made payable to Imprint Academic to PO Box 1, Thorverton, EX5 5YX, UK, or credit card details (Visa/Mastercard/Amex), including card expiry date. For more information contact Sandra Good. sandra@imprint.co.uk

Price: Individual $63 / £40.50. Institutional: $121 / £78
50% discount on complete runs of back volumes.
Sample copies are available on request.

Editor: Søren Brier, Royal Veterinary and Agricultural University, Dept. of Economics and Natural Resources, Section for Learning and Interdisciplinary methods, Copenhagen.

Mail address: Søren Brier, Landbohøjskolen, Inst. for Økonomi, Skov og Landskab, Sektion for Læring og Tværvidenskabelige Metoder, Rolighedsvej 23, st. th., DK-1958 Frederiksberg C, Denmark. sbr@kvl.dk
Telephone: +45 35282689, Fax: +45 35283709
http://www.flec.kvl.dk/sbr/index.htm

Lay out and art editor: Bruno Kjær, Royal School of Library and Information Science

Journal homepage: www.imprint-academic.com/C&HK
Full text: www.ingenta.com/journals/browse/imp

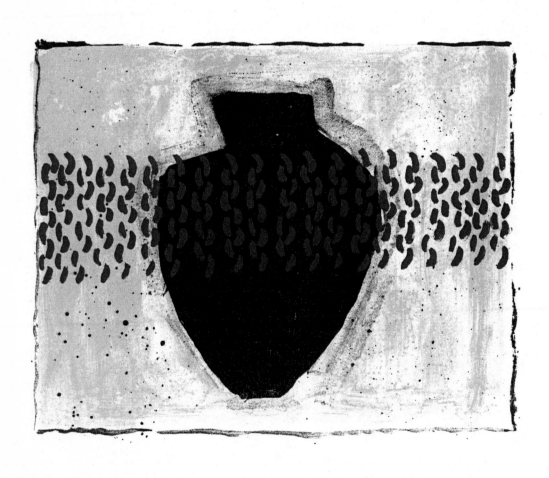

Cybernetics & Human Knowing, Vol.9, No.2, 2002, pp. 5–8

Foreword:
The Ages of Francisco Varela

Jeanette Bopry and Søren Brier

The present issue is a memorial issue for Francisco Varela both as a scholar and as a colleague. Varela passed away in his home in Paris on May 28 2001. He was part of the editorial board of this journal and thus in this memorial issue we would like to look into his heritage. Most of the papers we present have authors that have known and worked with Varela in some period of their and his life: Ranulph Glanville, Louis Kauffman, Andreas Weber.

Weber makes the case that Varela's thinking can provide a foundation for biosemiotics and as such it provides a further foundation for the cybersemiotic project. Most interesting and promising is his comparison with Varela's concept of the organism and Bruno Latour's concept of quasi-objects. The other articles all have some relationship to Varela's elaboration on the work of Spencer-Brown. Using the metaphor of the Uroboros, Marks-Tarlow, Robertson, and Combs explore the notion of re-entry in Varela's 'A Calculus for Self-Reference' and his contribution to a theory of consciousness. In their articles, Glanville and Kauffman reflect upon their experience working with Varela on joint papers. Glanville's paper is focused on self-reference and the Möbius metaphor, Kauffman's on Spencer-Brown's *Laws of Form*. The ASC column is by Søren Brier, who makes the case that Varela contributed to the bridge between second order cybernetics, autopoiesis theory and the triadic semiotics of C. S. Peirce. The issue concludes with a multi-authored historically-reflected review of *The Embodied Mind* (Varela, Thompson and Rosch) and its influences, followed by an afterword from Klaus Krippendorff.

Francisco's widow, Amy Cohen, tells us that his favourite poet was the 2nd century Buddhist monk, Nagarjuna. We are delighted therefore to include Stephen Batchelor's translation of Nagarjuna's poem 'Self'. This is particularly apt since Francisco was present at the first public reading of Batchelor's translation, at Dartington, Devon, in the UK. The other poems are contributed by Pille Bunnell and the artwork by Bruno Kjær.

From an ASC point of view Varela's contribution to the theory of autopoiesis with his teacher and mentor Humberto Maturana is one of the most important parts of his work. Heinz von Foerster supported Varela's development of Spencer-Brown's work in his article 'A Calculus for Self-Reference'. But Varela's intellectual history and influence goes both much deeper and broader than this. A self-described epistemologist, he was concerned with basic problems. The first of these was developing a definition of life. In his work with Humberto

Maturana, 'Autopoiesis' (in the book *Autopoiesis and Cognition*), they define the living system as an organizationally closed system of self-production, and cognition as the operation of that system. This theory has been fundamental to developments in the fields of biology, neuroscience and immunology. Further, autopoiesis theory and Varela's extension of it into autonomous systems theory (*Principles of Biological Autonomy*) has inspired many to reconsider their perspectives on their own fields; those in fields as disparate as family counseling, sociology, and law have been influenced by this work.

The idea of enaction theory, that we bring forth the world in the process of interaction, is an attempt to solve another basic problem: the subject/object dualism that plagues cognitive science. No doubt his development of the enaction theory has to be considered the crowning achievement of his career. It has had tremendous influence in the cognitive biology area. Jeanette Bopry in Vol.8 no. 4 of this journal used it most recently. The concept of embodied cognition provides, among other things, a means of accounting for common sense, something not easily dealt with in information processing approaches to cognitive science. The implications of these ideas are important for educators as well as cognitive scientists.

As an epistemologist Varela was concerned not only with forwarding our understanding of knowing, but also with the ethical consequences of knowing, as he expresses in the concluding paragraph of 'Whence Perceptual Meaning':

> In this essay, I have argue[d] that, if the kingpin of cognition is its capacity for bringing forth meaning, then information is not pre-established as a given order, but it amounts to regularities that emerge from the cognitive activities themselves. It is this re-framing that has multiple ethical consequences....to the extent that we move from an abstract to a fully embodied view of knowledge, facts and values become *inseparable*. To know is to evaluate through our living, in a creative circularity. (In Varela & Dupuy, 1992, *Understanding Origins*, p. 260)

His attention to values and ethics is explicated further in the text *Ethical Know-How*. In the past decade it became apparent that Varela's work had converged with his practice of Buddhist meditation and study of Buddhist philosophy. The importance of this orientation for him is especially apparent in the texts *The Embodied Mind,* co-authored with Evan Thompson and Eleanor Rosch, and *Sleeping, Dreaming, and Dying*, an account of a meeting between the Dalai Lama and prominent western cognitive scientists in 1992. In recent publications relating to neurophenomenology he proposes that Buddhist meditation practice provides one of the three methodological keys to a reliable phenomenological approach for accessing experience. His exploration of the connections between Buddhist practice and cognitive science directs us toward a future in which another basic problem will likely solved: understanding consciousness. The question of particular importance to him in consciousness studies is how cognitive entities experience temporality. For him, this is a question that can be explored both phenomenologically and experimentally. It is a hallmark of his career that he considered it important to engage in both experimental and reflective practice.

In John Brockman's book *The Third Culture* Varela is identified as a contributor to the Third Culture. He is among those scientists that philosophize over the meaning of new findings in the natural sciences and the transdisciplinary significance of the paradigm of complexity, self-organization and evolution. Stuart Kauffman points out in this book how amazingly inventive, freewheeling and creative Varela was combined with his smart, charming and graceful style. Kauffman also points out the connection between the autopoiesis theory and his own autocatalytic-polymer-set story. He sees a connection between Varela's work on immune networks and the Santa Fe Institute's results on emergent collective phenomena. Kauffman has some of the same beliefs about deep biological laws as Varela and discusses why these views have had so little impact on mainstream biology. Unlike physics and chemistry, biology is still not theory- and concept-driven. Biology is basically experiment-driven, as organisms, since Darwin, are mostly seen as ad hoc solutions to complicated design problems. Most biologists see Maturana and Varela as philosophers. As Danny Hillis points out in the same book most biologists, along with artificial intelligence and artificial life researchers, tend to view philosophy as a black hole that many a good scientist has fallen into – lost to real science forever! Some, like Christopher Langton, view autopoiesis theory as another way to map the same problems in biology that the mainstream is working with, without adding anything really new. Lynn Margulis acknowledges that the theory of autopiesis is useful as a means of distinguishing between living and non-living systems but, with many other experimental biologists, has problems following the more philosophical aspects of Varela's ideas. Daniel Dennett comments that he has been inspired by Varela but finds him too revolutionary and confusingly Buddhist. From the mid-seventies Varela practiced Tibetan Buddhist meditation combined with studies of Buddhist psychology and philosophy. He hoped that this spiritual and existential tradition and Western cognitive science could benefit from each other. In *The Third Culture*, the more philosophical biologist Brian Goodwin actually praises Varela's ability to funnel his theories into high quality experimental research.

This discussion says something about the difficulties that even the most gifted among researchers have in embracing other views of their field. They are able to take in the more concrete aspects of Varela's thinking, but the whole picture is too much to embrace – too revolutionary to be broadly accepted in his own generation. But time and the advent of new generations brought up in more philosophically reflective forms of science will work for the acceptance of his grand vision, including Western phenomenology, as we can see in the articles in the present issue.

Francisco Varela was not only a pathfinder, but also someone who saw and inspired others to see how seemingly disparate paths could be connected. Those of us interested in interdisciplinary fields like cybernetics, cognitive science and semiotics celebrate this contribution. We can only imagine what other paths this creative intellect might have illuminated for us, and this amplifies the loss we feel. Below we provide a few selected references. For more details about Varela's life and work see Thompson (2001).

Selected References

Maturana, H.R. & Varela, F.J. (1980). *Autopoiesis and cognition: The realization of the living.* Boston Studies in the Philosophy of Science, vol. 42. Dordrecht: D. Reidel.

Petitot, J., Varela, F., Pachoud, B., & Roy, J.M. (Eds.) (1999). *Naturalizing phenomenology: Contemporary issues in phenomenology and cognitive science.* Stanford, CA: Stanford University Press.

Thompson, E. (2001). Francisco J. Varela (1946-2001): Tribute by Evan Thompson." *Journal of Consciousness Studies, 8*(8), 66-69.

Varela, F.J. (1966a). The early days of autopoiesis: Heinz and Chile. *Systems Research, 13*(3), 407-416.

Varela, F.J. (1966b). Neurophenomenology: A methodological remedy for the hard problem. *Journal of Consciousness Studies, 3*(4), 330-350.

Varela, F.J. (1979). *Principles of biological autonomy.* New York: Elsevier North Holland.

Varela, F.J. (Ed.) (1997). *Sleeping, dreaming, and dying: An exploration of consciousness with the Dalai Lama.* Boston: Wisdom Publications.

Varela, F.J. (1999). *Ethical know-how: Action, wisdom, and cognition.* Stanford, CA: Stanford University Press.

Varela, F.J. (2001). Intimate distances: Fragments for a phenomenology of organ transplantation. *Journal of Consciousness Studies, 8* (5-7), 259-271. Reprinted in E. Thompson, ed., *Between Ourselves: Second-person issues in the study of consciousness.* Thorverton: Imprint Academic (2001), [see back cover, *C&HK,* this issue].

Varela, F.J. & Bourgine, P. (Eds.) (1991). *Toward a practice of autonomous systems: Proceedings of the First European Conference on Artificial Life.* Cambridge, MA: The MIT Press.

Varela, F.J. & Coutinho, A. (1991). Second generation immune networks. *Immunology Today, 12,* 159-166.

Varela, F.J. & Dupuy, J.P. (Eds.) (1992). *Understanding origins: Scientific ideas on the origin of life, mind, and society.* Boston Studies in the Philosophy of Science, vol. 130. Dordrecht: Kluwer Academic.

Varela, F.J., Lachaux, J.P., Rodriguez, E., & Martinerie, J. (2001). The brainweb: Phase synchronization and large-scale integration. *Nature Reviews Neuroscience, 2,* 229-239.

Varela, F.J. & Shear, J. (Eds.) (1999). *The view from within: First-person approaches to the study of consciousness.* Thorverton: Imprint Academic.

Varela, F.J., Thompson, E., & Rosch, E. (1991). *The embodied mind: Cognitive science and human experience.* Cambridge, MA: The MIT Press.

Total publication list: http://www.ccr.jussieu.fr/varela/index.html

* * *

The president of the ASC, Pille Bunnell, has donated the poem on the facing page with the following dedication: "To Francisco Varela, who I never met in person, but whose touch I see in many friends and thus I know him. May his goodness live on!"

questions

Someone asks
and I see
what I did not see before.

Did I know that?
Did I simply omit to look?

No.
My understanding arose
with the question.

A true question does not
demand an answer.
It is not a hypocrisy
with the nature of the answer
preconceived.
A true question
is an honest question.

An honest question
arises in love,
in the acceptance of a world
where what appears
is legitimate.

An honest question is a request
for an expansion of understanding.
In this desire illumination happens
and I see
what I could not see before.

This is a gift
from the one
who asks.

Cybernetics & Human Knowing, Vol.9, No.2, 2002, pp. 11–29

The 'Surplus of Meaning'. Biosemiotic aspects in Francisco J. Varela's philosophy of cognition

Andreas Weber[1]

Abstract: The late Chile born biologist Francisco J. Varela has been influential in theoretical biology throughout the last three decades of the 20. century. His thinking shows a marked development from a biologically founded constructivism (developed together with his fellow citizen, Humberto Maturana, with the main key word being "autopoiesis theory") to a more phenomenological oriented standpoint, which Varela called himself the philosophy of embodiment, or "enactivism". In this paper, I want to show that major arguments in this latter position can be fruitful for a biosemiotic approach to organism. Varela himself already applies concepts as e.g. "signification", "relevance", "meaning" which are *de facto* biosemiotic. He derives these concepts from a compact theory of organism, which he understands as the process of self-realization of a materially embodied subject. This presumption stems, though somewhat modified, from Autopoiesis theory and so attempts a quasi-empirical description of the living in terms of self-organisation. Varela's thinking might count as an exemplary model for a biosemiotic approach in a theory of organism. In particular, Varela's link to down-to-earth biological research offers means to associate biosemiotics with the ongoing debate about the status of a biological system within genetics and proteomics research.

"I want to start declaring that I think that understanding of organisms and the living is possible, that defining these terms in a satisfactory manner is not a utopian dream, and that we even have a good deal of the road already charted. However, this is under a fundamental condition: that the autonomy of the living is highlighted instead of forgotten, as it has been."[2]

1. Introduction

It seems crucial for the further development of biosemiotics, to be grounded most tightly in a solid biological account. What is required, are empirical observations of how self-organising systems are bringing forth signs, and how they make work the molecular context of the body in semiotic terms.

The philosophy of Francisco J. Varela provides an important source for such an undertaking. His work has never lost its contact to empirical experiment. It is a bridge between research in biological cognition, and a completely original account of what the phenomenon of life is. Most interestingly, some of Varela's most influential contributions to this field are already expressed in biosemiotic or at least "cryptosemiotic" (see Sebeok, 1976, for this expression) terms, apart from

[1] Institute for Cultural Studies, Humboldt-Universität zu Berlin, Sophienstraße 22a, D–10178 Berlin, Germany, e-mail: andreas.weber@rz.hu-berlin.de. Private Address, which should be used for correspondence: D-21720 Guderhandviertel 90.

[2] Varela, F.J. (1997): "Patterns of life: Intertwining identity and cognition". *Brain and Cognition* 34, S. 73.

other, more phenomenological points which stand in a loser contact to biosemiotics proper and hence are not subject of this essay.

Francisco Varela's work might help explain semiotics from the basal fact of organic life making. It could provide a possible way to understand the transition from a purely cognitive level to a semiotic one. Varela's description of the "patterns of life" is in fact the reconstruction of a semiotic nucleus. In such a view, semiotics is coexistent with life, because life always is embodied cognition, giving rise to a world of relevance. However, Varela himself did not absolutely share these consequences. He was always reluctant about semiotics, seen the relatively formal design, which the discipline acquired under the influence of Eco (e.g. 1976). *Avant la lettre*, one might thus say, Varela has written a biological foundation of (bio-)semiotics. His work shows a close association of the fields of embodied cognition, biological phenomenology, and semiotics, conflating in what could best be called a biology of subjective agents, or short, *subject-biology*.

2. Biological constructivism and beyond

Varela's ideas about organism set off from the concept of autopoiesis, which he developed as a young scholar together with Humberto R. Maturana. Since then it has mostly be known under the label of "biological constructivism". I will discuss these roots here in some detail. This is necessary because later Varela has much shifted accents in his view on organism. He tried to escape some of the less desirable consequences of his early theories, a parting that gave rise to the intellectual separation of the former co-authors. Maturana stayed on a more mentalist, less embodied account of cognition (Maturana 1987), whereas Varela moved to what he later called "enactivism" and which is coming much closer to biosemiotics.

Biological constructivism, which Varela and Maturana developed in the 1970s (Maturana & Varela 1980), was an attempt to overcome the philosophical fission of the phenomenal world. The authors tried to conquer the dualistic gap between objects and perceivers, or things and subjects, that has haunted philosophy since early modern times. Instead of a neat separation of perceiver and perceived, Maturana and Varela argued that the structure of favouring phenomenal worlds is created by the biological structure of an agent that has to make a living in this world.

In biological theory, dualism means the dominant understanding of the living as molecular-genetic and, complementary, the process of mind and cognition as information processing. What is missing, for Maturana and Varela seemed to be the fact that an agent realizes the living actively. An organism is an autonomous being that does not encounter the world passively, but experiences it as a meaning. Maturana and Varela hence tried to synthesize what could be said to be the minimal form of autonomy consistent with modern cell biology, with the emerging studies on self-organization and systems research circulating at the time (von Foerster, Rosenblatt, McCulloch, Wiener) [3].

The result was autopoiesis theory. A living being is an autopoietic entity, which is characterised by the fact that it is literally producing itself (*autopoiesis* is derived from the Greek, meaning "self-creation"). The particular character of the autopoietic system is that the cell's metabolism brings forth entities that will feed back materially, i.e. integrate into the network of transformations which has brought them forth (Maturana & Varela 1987:50).

A living system hence creates itself and its components in certain autonomy from its surrounding world, and does so even against external gradients. Differing from dissipative model systems and from all other self-organising reactions (and, as we will see later, also from self-regulating artifacts), organisms are defined by their active behaviour in searching food, escaping danger and generally keeping up their integrity against external influences.

Regarding cognition, the active self-realization means that an organism is not concerned with any external events, but only with the circular process of bringing forth itself. The external world is hence constructed by means of the inner, biological structure of the living agent. This holds true as a principle also for higher levels of integration: not only the cellular metabolism, but also the nervous system is closed. It brings forth only its own inner states, which can be stimulated by the environment, but may not be influenced causally in an unambiguous way. The nervous system does not receive "information". It rather creates a world by defining which configurations of the milieu are stimuli (also called, as we will see below, "perturbations"). This interpretative world making is the only way, which can provide "information". Autopoietic agents create a world by their actions (Maturana & Varela 1987:185)

This solution was a neat reformulation of Kant's transcendental philosophy in the guise of an empirical theory of cognition. Subjects perceive the world by means of phenomena, "Erscheinungen", which are determined by the biological (and not longer, as in Kant, transcendental) conditions of being a subject. Biological constructivism hence can be said to put Kant onto a biological foundation and to corroborate his transcendental philosophy by means of natural sciences. The reality is constrained by the biological conditions of any possible knowledge. World is not objective but relative to the biological organisation of the perceiver.

The problem here, which led to a crisis in autopoietic theory, was already a difficulty for Kant. If subjects bring forth their own cognitive worlds, how can we avoid solipsism? How can communication with fellow beings be possible, which obviously is a most central fact of life? In the first presentations by Maturana and Varela and in Maturana's later work, it seems of no importance how the external world is structured, and if there is *something* at all. Thus, Maturana and Varela (1987:149ff) come to compare the situation of the perceiver to that of a

[3] For historical background information see Varela 1996.

submarine-driver who relates to the external world only by means of his navigation equipment. He experiences reality only as a technical translation, a construction achieved by means of his instruments. Reality here could as well be the fiction of an evil demon fed into the instrument's sensors. The subject is unable to act back on its surrounding world. We can see that by this construction, dualism, which autopoiesis theory was conceived to weaken, is rather reinforced (see also Maturana, 1987:89-118).

Varela in his more recent works does no longer so much emphasize the hermetical dimension of autopoietic systems. An important facet of a living being's reality still remains in the fact that this reality is created by the being's own world making. However, on Varela's "middle way of knowing" between constructivism and representationism, the world as a material and energetic ground is always present. It is even vitally necessary as the source for an existential coupling, which specifies both agent and Umwelt. Instead of emphasizing construction, Varela later prefers to speak of enacting, of a "mise en scène" (Varela 1997). The resulting philosophical position is "enactivism" instead of constructivism, thus confronting the crucial question of "the other" (Varela et al. 1991) that only few positions in continental philosophy have dwelled upon so far. In the following, I will be summarizing this new position.

However, let me observe before that by this turn to the "exterior", in the work of Varela we can witness the same movement, which also Kant finally took in his life's work. Kant had all the time tried to reconcile the empirical variety of experience (and of objects) with the existence of a transcendental (we would say: constructivist) subject. But only in the third of his critiques, in the Critique of Judgement, he achieved a kind of preliminary unity between subject and experience, the link, which he was so eagerly looking for. But to be consistent with the premises of his transcendental, "un-embodied" critical position, Kant had to pay the price of admitting a "happy chance" in the fitting of human perception to the world of objects (Lenoir 1982).

In the second half of the Critique of Judgement, Kant is concerned with organism and its apparently teleological organisation. To reconcile transcendental subject and experience, it seemed necessary for Kant, to admit a teleological description of organism, even if it contradicted the findings of the earlier critiques. Thus, Kant elaborated his somewhat ambiguous and indecisive position about teleology in biology, which has aroused so much confusion until today. But, in relation to autopoiesis theory, teleology might be seen in a new light. As Varela and Weber (2002) have shown, the enactive turn of autopoiesis can provide an explanation for "intrinsic teleology" and so strengthen out Kant's position to this side. We then have in autopoiesis, by its intrinsic teleology, along with its constructivism, a double naturalization of Kant.

As we will see below, this explanation of teleology from the standpoint of an embodied phenomenology, can in turn be transferred to a biosemiotic account. As already Peirce showed, both semiotics and teleology are triadic forms of relations (Weber 2001a, but see also below).

3. The process of establishing an identity

For Varela, organisms enact the world they are living within by the process of their auto-constitution. In the very process of self-creation they put onto stage their proper—and in these terms their real—world. With this in mind the shortest definition of organism is: "Organisms are fundamentally a process of constitution of an identity." (Varela 1997)

For autopoiesis theory, the process of the living consists in bringing forth this proper process. Living is ended not when its chemical compounds are changed, but when the process of auto-maintenance is disturbed. What does this auto-creation of a unity mean on a concrete biological level? Varela characterises living beings by the fact that they literally produce themselves. An organism brings forth its structures as well as its own border. It is the production of an order which produces exactly the components that have produced the order:

> An autopoietic system—the minimal living organization—is one that continuously produces the components that specify it, while at the same time realizing it (the system) as a concrete unity in space and time, which makes the network of production of components possible. More precisely defined: An autopoietic system is organized (defined as unity) as a network of processes of production (synthesis and destruction) of components such that these components (i) continuously regenerate the network that is producing them, and (ii) constitute the system as a distinguishable unity in the domain in which they exist. (1997:75)

It is important to keep hold of the detail that the living entity exists as a certain identical structure in space and time, although at the same time it is at no moment materially identical with itself. The living entity thus has a strangely virtual aspect; the actual material being is in a way the product of this virtual facet, although both are in any moment also identical. Only the fact of being alive keeps this circuit closed. When an organism dies the process comes to an end, the components behave as normal chemical compounds and decay.

It should be clear that the distinguishing characteristic of a living being is its nature as a process or even a network of processes. An organism brings forth its own structure and limits itself as a distinct entity. Its identity consists thereby exactly in bringing forth this identical 'self'. Without the process of this steady auto-production, the entity would immediately collapse.

Obviously, the organization of the living is characterized by a strong circularity and a conflation of two rather different ontological realms. The organism's proper reality unfolds as a dialectics of pure matter and structural regulation. This process circularity is a fact, which biology has discovered in the field of complexity science emerging since the 1960s and with today protagonists as Kauffman, Goodwin, Rose and Salthe. There is nothing vitalist in the view of emerging circularities; even many chemical reactions display self-organizing behaviour (Prigogine & Stengers 1990). The process of the living happens on normal matter—only that this is organized in such a way that it comes to autopoietic behaviour.

(Findings like these call for a more detailed investigation in the direction Weber and Depew (1996) pointed out. They proposed the existence of a large framework of structural rather than causal laws in biology, causal models as Darwinian evolution being rather a subclass of a deeper structural grid providing certain creodes of self organization.)

As a model autopoietic system, we can regard a prokaryotic cell. Although it is rather simple, it retains a certain incontestable ontological unity and shows already the full behaviour of a living being. Although model systems (as in Varela 1991) help to illustrate how Autopoiesis works, its essence can only be observed in a living cell. This is thus by far superior to any artificial 'minimal autopoietic system' as a philosophical object of investigation. Only the living cell is autonomous in its active uptaking of nutrients. It is not dependent on a experimental setting to continue. So far, such a system has not been simulated. What is distinguishing an organism from any simulation is this active concern about itself. Therefore, autopoiesis is not a system theory of biological entities in a common sense, nor a mere variant of self-organisation theory:

> In its original formulation as well as in the subsequent literature it has been common to see the central concept of autopoiesis as just one more self-organizing mechanism (which it undoubtedly is), and even to conflate it with dissipative structure or autocatalytic cycles, or mere open systems. These ideas basically stay within the perimeter of a physicalist view of nature and understand these new developments as necessary extension of classical physics. However there is a an essential difference between these views and autopoiesis: autopoiesis proposes an understanding of the radical transition to the existence of an individual, a relation of an organism with it-self, and the origin of "concern" based on its ongoing self-produced identity. One could envisage the circularity metabolism-membrane entirely from the outside (this is what most biochemists do). But this is not to deny that there is, at the same time, the instauration of a *point of view* provided by the self-construction. (Weber & Varela 2002)

By creating the self in a process of unfolding identity, the organism brings forth its surrounding world as the 'other' of its self. Between the organism and the world thus brought forth, exists a certain paradox. The living creates its surroundings, and consequently is rather autonomous. However, at the same time it is still depending on these surroundings as a material source of its real existence:

> Organism connotes a knotty dialectic: a living system makes itself into an entity distinct from its environment through a process that brings forth, through that very process, a world proper to the organism. (Varela 1991:79)

As a visible expression of this dialectical linkage (or "coupling"), organisms are contained within a material border: a cuticle, a skin, a shell etc., which is at the same time a product of the organism and its limit to the exterior. One could nearly say: organisms are *prima facie* an act of constant self-separation from the surroundings.

The world beyond this border is materially identical with the compounds that huddle together behind the fluctuating, and at the same time stable organic boundary. However, because the exterior has no importance for the autopoietic process, it is also different from the matter flowing through it. The organism defines what is outside and what is inside by its self-realization. The distinction thus is pragmatic and not fundamental; it is taken relatively to the system and thus subjectively from its perspective. At the same time though, the distinction is an absolute one: when the autopoietic process stops, the system immediately disintegrates to 'pure outside'.

This dependent independence is a decisive trait of the living. If the process of life is already a dialectical, not substantial, but rather relational affair, then this is the case far more prominently for the resulting phenomenology, which is brought forth by the living: Inside and outside are not *really* separated, *only* in relation to the process of self-establishment. Subject and object are not 'really' separated but become so only in the process of constitution of a subject *vis-à-vis* to its substrate. It is interesting that this insight reminds of the understanding of quantum mechanics, where in certain experiments subject and object cannot be separated entirely from one another and thus must be identical in a deeper sense.

4. Emergent levels of "Self"

The horizon against which this existential-semiotic activity is brought forth, only in simple unicellular beings is the autopoiesis of the *cell*. In other cases, self-referentiality appears on rather different emergent levels of self, which are linked via the circular closures that higher organisms have developed as a means of self-reassurance. But in principle the basic relation, a fragile embodied closure keeping up against *and* by means of a material substrate, is the same even on more complicated levels of biological selves. Hence, Varela distinguishes

> 1) a minimal or cellular unity, 2) a bodily self in its immunologic foundations, 3) a cognitive perceptuo-motor self associated to animal behaviour, 4) a socio-linguistic 'I'' of subjectivity, and 5) the collective social multi-individual totality. In all these regions we are dealing with levels and processes where an identity comes about—not as substance, but as movement—and whose fabric of organisation *is* the organism. (Varela 1991:80)

In the quotation, points (4) and (5) are more complicated and not related only to autopoiesis (I discuss this fact extensively in Weber, 2002b). Cultural or social semioses might mirror their organic deep layer in a semiotically quite refined form. This is the standpoint of Hans Jonas. He has developed a proto-autopoietic view of metabolism as paradigm of the existential situation every living being is embedded in. Jonas argues that even the paradoxes of human existence, which oscillates between autonomy and necessity, stem from the making up of organic identity by metabolism. (Jonas uses a slightly different terminology, see Weber 2002a). It might be possible that human ways of world making can (or even should) be traced back to a general organic layout.

What is necessary as a horizon for providing semioses, or, as Varela would have it, for achieving genuine cognition, is an embodied process of creating an identity, a process, which is not instructed by external means, and which is thus always open to failure:

> In defining what it is as a unity, in the very same movement it defines what remains exterior to it, that is to say, its surrounding environment. A closer examination also makes it evident that this exteriorisation can only be understood, so to speak, from the "inside": the autopoietic unity *creates a perspective* from which the exterior is one, which cannot be confused with the physical surroundings as they appear to us as observers [...] (Varela 1991:85]).

This definition obviously does not apply to artificial life so fare conceived, even if the computational paradigm has given way to the self-organisational view (Brooks 2001). A better way to create life would rather consist in developing a system that as a foremost goal seeks to maintain its own continuity (which might be a dangerous experiment). Self is brought forth only as a closure on and against an influence from a world without that is defined only by this closure—a highly circular process that at the same moment specifies the agent *and* the surrounding world.

The idea of a process of identity also has advantages over the common view, which explains organic identity by genetic similarity. Rather, it has become clear that functioning bodies, i.e. "existential units" can in no way be defined by genetic homogeneity, which has so much come into focus by actual work. Genes may be sequenced, but they do not account for the completeness of a whole organism:

> Identity is not an object; it is a process with addresses for all the different directions and dimensions in which it moves, and so it cannot so easily be fixed with a single number (Margulis und Guerrero 1991:50).

There are several points accounting for that. One is concerning the loss of parts of the genome during ontogenesis. Another, more important, deals with the phenomenon of symbionts living as a part of the bodies of most metazoan organisms. This factual inter-species heterogeneity appears as a serious problem for every essentialist theory of biological identity:

> Man appears as a complex to whose harmonious well-being many bacteria, for instance, are absolutely essential. Intestinal flora are needed for metabolism, and many kinds of bacteria living in the mucous membranes are required for the normal functioning of those membranes. (Fleck 1979 [1935]: 59-60)

It is of central importance for metazoans, that their cells have to communicate about how to make up the higher unity they are a part of. Most organisms consist of billions of individual cellular bodies and even of other, symbiotic organisms that finally make up the being we perceive as unity:

> The constituents are closely interdependent and on their own are usually not viable. All symbioses, for instance between nitrogen fixing bacteria and beans, between mycorrhiza and certain forest trees, between animals and photogenic bacteria, and between some wood beetles and fungi, form 'harmonious life units', as do animals, such as the ant colony and ecological units

such as a forest. A whole scale of complexes exist, which, depending on the purpose of the investigation, are regarded as biological individuals. For some investigations the cell is considered the individual, for others it is the syncytium, for still others a symbiosis or lately even an ecological complex. (Löwy 1991:43)

Symbiosis obviously is a much deeper biological regularity than long has been thought. Without it, probably no multicelluar organism would exist: already the higher (eukaryotic) cells probably arose by fusion of different bacterial cells. Every organism is vitally dependent on its symbionts and vice versa, without their mutual interaction both would perish. Obviously, the symbionts change their host in such a way that it can only exist as such *if changed*: The clear separation of host and symbiont holds no longer true. As observe Chernyak & Tauber (1991:129):

> From this perspective, the concept of individuality does not coincide with genetic homogeneity. Thus, when various strategies are reviewed, the biological lesson concerning homogenous individuals is that *individuality* may not be defined as genetic 'sameness'.

Only by taking a functional criterion into account, as is the establishment of the self by a process of identity, we can avoid substantial notions in defining what an organism is. To speak of a process of identity, however, inevitably introduces the semiotic standpoint.

5. Self-constructing the semiotic perspective

The autonomous construction of reality, done by a living agent, Varela calls cognition (using a terminology in the tradition of Biological Constructivism). In cognition, organisms bring forth a world: they create relevance by separating the outside from their selves and at the same time remaining dependent on the outside as a source for existence. Cognition in Varela's sense is already a semiotic concept. In his term "surplus of significance" (see also below) Varela is providing a semiotic theory *in nuce*. Significance, or meaning, associate with the act of continued existence, and spring forth as a basic feature of a living being's world making:

> In brief, the term cognitive has two constitutive dimensions: first its coupling dimension, that is, a link with its environment allowing for its continuity as individual entity; second its *interpretative dimensions*, that is, the surplus of significance a physical interaction acquires due to the perspective provided by the global action of the organism. (Varela 1997:81, emphasis by me, A.W.)

In his 1991 paper, Varela even uses the expression "imaginary dimension", a keen term, but maybe better characterizing that what we could understand by 'surplus', as we will see now. The changeover from the material level to the level of significance happens by the same movement by which the organism manifests as a global process, in other words, when there is life. For an organism, any relation of energetic trade with the milieu becomes a relation of signification on the existential setting of the organism. The formation of the self creates the surrounding world on the background of this exchange process. In the act of self-assurance self and foreign are (pragmatically) defined. Because what has been exteriorised by the system as foreign is still vitally important for the it (as food, shelter, even mate), the system's domain of relevance is created. The relevant world is the system's *Umwelt* established by this process:

> It is *ex-hypothesis* evident that an autopoietic system depends on its physico-chemical milieu for its conservation as a separate entity, otherwise it would dissolve back into it. Whence the intriguing paradoxality proper to an autonomous identity: the living system must distinguish itself from its environment, while *at the same time* maintaining its coupling; this linkage cannot be detached since it is against this very environment from which the organism arises, comes forth (Varela 1991:85)

The dialectic involved here is a quite fundamental one: Only by self-definition, the world as the domain of relevance is established, and only in the separation from this world, the self supplies signification. By confrontation with the world, which it has excluded actively, the subject generates meaning. Thus the

> cognitive activity is paradoxical at its very root. On the one hand, the action that brings forth a world is an attempt to re-establish a coupling with an environment, which defines the internal coherence through encounters and perturbations. However, such actions, at the same time, demarcate and separate the system from that environment, giving rise to a distinct world. (Varela 1997:80)

Autopoiesis thus differs from other self-organizing concepts in that it is on one hand close to strictly empirical grounds, yet on the other hand provides the decisive entry point into the origin of individuality and identity, connecting it hence with the semiotic realm.

Varela understands the organism as a relation between compounds and the process creating these compounds, and thus as a relation between a self and an *Umwelt* together with this self. This latter relation is in a strong sense existential, because it belongs to the network of processes by which the organism distinguishes itself from matter; it is literally deciding about prosperity or failure, about life or death. This existentiality branches out into the finest filiations of possible relations to the world. And it is the organism's paradoxical dependence on its surrounding that lends to that surrounding an unchangeable existential meaning. On this foundation, a semiotics of the living world can be constructed. Its shape would be a biological theory of meaningful natural signs *qua* the organisation of the living that is realizing its existence.

6. Need, value and meaning

Meaning arises as a kind of ontological foil for the material auto-production, which an organism is concerned with *ad infinitum*. Therefore, subjectivity is not only found in human intentionality. It is rather at the ground of any behaviour emerging from the autopoietic outset. Subjectivity is the expression of the fact that a living system is concerned with itself. *Because* life is continued existence against the weight of matter, life is already intentional at its very beginnings, and there is a subject perspective in every living system. This subjective perspective is the standpoint of concern: a living system tries to keep up itself against external influences, or, as we could say in autopoietic terms, perturbations. Already basic forms of life, therefore, adopt a subjective perspective as a result of their existential need.

Life is a fragile, precarious principle. Life is not an unlimited success-story, because it is a process happening on substantial matter. The self-making self has to survive in a world characterised by an "other". In the antinomy of form and matter, which lies in the metabolic principle of life itself, this other-reference becomes a first order phenomenon. The rudimentary cognitive subject is the *interpretant* necessary to make up a biosemiotic entity:

> The difference between environment and world is the surplus of *signification* which haunts the understanding of living and of cognition, and which is at the root of how the self becomes one... There is no food significance in sucrose except when a bacterium swims up gradient and its metabolism uses the molecule in a way that allows its identity to continue. This surplus is obviously not indifferent to the regularities and texture (i.e. the "laws") that operate in the environment, that sucrose can create a gradient and traverse a cell membrane, and so on. On the contrary, the system's world is build *on* these regularities, which is what assures that it can maintain its coupling at all times. (Varela 1991:86)

Only the presence of a living being provides the objects in the world with their sense. It does so by transforming them to the stage set of an existential drama. Only by this existential sense, objects gain their significant role. Their presence or absence for the organism decides about prosperity or defeat, stability or chaos. This happens precisely because the organisms has to "master things", because they are so detached from its autonomy that they only *mean* something to him and do not *cause* a behaviour directly. As Varela observes:

> Whatever is encountered must be valued one way or another—like, dislike, ignore—and acted on some way or another—attraction, rejection, neutrality. This basic assessment is inseparable from the way in which the coupling event encounters a functioning perceptual-motor unit, and it gives rise to an *intention* (I am tempted to say "desire"), that unique quality of living cognition (Varela 1991:97).

The subjective world of existential meanings arises in the same movement by which the organism is creating itself. It is thus necessarily coupled to such an auto-creation and cannot be separated from it. Rather, when organisms are conceived of as autopoietic systems, *meaning* is one of their fundamental dimensions of existence—it is the true marking point that distinguishes the organic realm from matter.

The existential dependence that is the consequence of the special mode of being of the organism is a motor of meaning *ex negativo*. The endless swelling of the imaginative realm, of the domain of options and of creation is the other face of lack that the living every time has to cope with to go on (for a inspired discussion of the ontology of genuine creation cf. Steiner 2001). It is therefore not so badly exaggerated when Varela (1991) speaks of an "imaginary dimension" of the living: the answer to the dependence is the unfolding of a dimension of meaning. In all its creativity the living uncouples from the crude existential situation, and nonetheless always keeps related to it. This relation is a play with the constraints in an endless morphological, esthetical and maybe, moral variation. The perspective of a threatened and thus affirmation-dependent organism lays a new grid over the world: a ubiquitous scale of value. Everything the living being is interacting with, gains its own value in the pragmatics of this interaction. Its relevance is related to the amount in which it allows the continuity of existence. For an organism the world is at no a time neutral place.

The entire colourful ontological universe we know arises from this perspective. It can only unfold in a fragile being that is all the time threatened by its destruction and thus invents, or enacts, ever higher levels of integration. The world without living agents would be a completely neutral place. Only after life has come into it, the world becomes real in prospering and pain, joy and misery. Only the living is interested in its life as continuity. By this interest it introduces an absolute value into the indifferent matter. This *absolute meaning* is the only reliable constant in an organism's life.

Things are not neutral in them, but already marked by good or bad. In the first beginning, they have no names except the existential notion of help- or harmful. Therefore, we could say with Jonas (1973) that *feeling* is the first unfolding of the world, and maybe we find an emotive background as the deepest underlying structure in all concepts of reality (Weber 2002a). The first fission of the world, the first discontinuity in the homogenous equilibrium of eternity has no structure, it is nothing but the amorphous cry of highest urgency uttered by the organism: vital or deadly.

This point provides strong affinities to two adjacent discourses. On the one hand, from a biological perspective, Kull (2000) has highlighted that ontologically we can understand the living, as opposed to functional artefacts, as displaying "need". From a phenomenological perspective, Barbaras (1999) has proposed to comprehend the living and its intentional worlds as being grounded in a deep layer, which he calls "desire". This is the background that *ex-negativo* forces biological beings to invent their perceptive and morphological worlds as a positive movement to evade stagnation viz. destruction.

However, this negative motivation must not necessarily lead to a *negative ontology*. As there is no relevant world *before* the organism, there is no threat to its closure before it is achieved. Desire is the complement to the exterior world, which the organism creates by its auto-affirmation; the lack is overcome or even over-satisfied at the moment of its occurrence. Rich variety and experience immediately fall together with the possibility of their destruction. Before semiosis, only meaningless gradients exist.

7. Intrinsic teleology: perturbations become signs

It is interesting that the complex network of a living being is able to quit the linearity of cause and effect. Its properties as a network, which are empirically observable, stand in a clear opposition to the attempt of reducing biological processes to a linear genetic causal principle. Genes thus can be viewed as one part of a larger regulatory context. A system, which is not *reacting* directly to an input, but *acting* according to its inner disposition, is not strictly bound to mechanical causality. However, this autonomy is not achieved by the introduction of new "vitalist" laws of live, but by the emergent regime that the interacting compounds produce. Thereby the outer world becomes a parameter in the complex self-regulation. The immune system may be a good example on the level of a functional. Also Varela (1991) devotes detailed work on it (see also Varela & Anspach 1991).

Important philosophical problems, explicitly those of free will, exist only if the organism is viewed in terms of a deterministic machine and not as an autonomous system. The Varela school is emphasizing that the external world acts as a mere 'kick', which motivates the system to establish a new equilibrium characterised only by the necessities of self-support. For a biosemiotic approach this means that it is not longer concerned with the constraints of the mind-body-problem. Dualism becomes obsolete by the material circularity of autopoiesis. In a self-referential system, meaning is the "inner" side of the material aspect of the system's closure. Dualism thus is swapped with a semiotic standpoint. Consequently, a biosemiotic account must show the transition of meaning from its material origin to its subjective meaning in organism. On a more phylogenetic basis, it is concerned with the evolution of sign-processes from a general organic level to a human or cultural level (for some ideas on this see Weber 2001a, 2002b).

At this point new light is shed on the closeness of the concept of semiosis to the idea of teleology. As we have seen in chapter 2, autopoiesis promotes a realist reading of "internal" teleology: why can organisms function purposefully? What is not touched by it is the notion of "external" teleology, the question, if the biological world is a product of purposeful design, which occupied physicotheology and motivated Darwin's thinking in the beginnings (See Weber & Varela (2002) for a detailed argument in favour of a reconsideration of teleology as a founding condition of the living.)

To restate it briefly: An autopoietic description amounts to a realist reading of intrinsic teleology and at the same time can be expressed in semiotic terms. Already Peirce saw this close relation of teleology and semiotics, as Deely (1990:84) observes: "To Peirce, the fact that a sign always signifies something to or for another suggested the need to reconsider the taboo notion of final causality, or so-called teleology". The making of the self actually creates a triadic situation: by (1) self-confirmation, the (2) non-self as *Umwelt* is separated, and this separation now opens (3) the option for interactions with the *Umwelt* in the better or worse. On one hand, this triadic relation can be understood as the archetype for a sign-process in the sense of Peirce. On the other, it is the way along which real world making takes place in organism. Living beings are embodied teleological processes. Intrinsic teleology calls for the description of organisms as subjects. Intrinsic teleology can be described in semiotic terms.

Hence, in biology, we have already an embodied paradigm for semiotic interactions. Sense making is a real, existential activity of organisms: Still more, it is of existential importance and even can be defined as the basic character of organisms. Life means to make sense. Self-constitution of a subject *always* is the constitution of a semiosphere. This is the necessary condition to write a biological semiotics as a foundation for a general semiotics.

Peirce's triadic sign has another trait, which we can find in Maturana's and Varela's (1987:179ff) reasoning. They have created the term "perturbation" as a description of the system's coupling to the surrounding world. A perturbation is a (indifferent) stimulus that is interpreted by the living being according to its inner structure. A perturbation becomes a sign through its meaning for the ongoing existence of the organism. From an internal perspective, every perturbation is experienced as a sign. Only the sign exists really (in an energetic sense). However, it is but perceived in that form which our body makes of the perturbation it causes. Conversely, the interpretant is real only in so far the sign has a meaning for him.

The object (in our case: the signification) does not 'really' exist as such, but only arises as it is touched by energetic-material influences of the surrounding world. Such a conception again bears similarities to Kant's *Ding an sich*, which can never occur in its "real" nature because we are slaves to our conditions of perception. When we see this closeness of Kant to a semiotic interpretation, we can understand why Peirce as well as Uexküll felt they would continue the Königsberg philosopher's argument.

In a pragmatic sense, the only reality is signification: it springs from an existential need and thus becomes the elementary grid of experience. Experienced meaning thus is the only trace of an underlying, complicated ontology. In the organism's perspective, nothing remains except from this signification. The existential condition is meaning, and while being existential, it hides its ontological foundation and thus becomes the proverbial blind spot on the organism's standpoint.

8. The living world: co-specification and interbeing

As we have seen, Varela embeds his theorizing in a general model of organic cognition, i.e. in a theory about construction of meaningful phenomenal worlds. His later work therefore shows a strong occupation with the question how the closed semiotic universe of an organic agent can be compatible with the notion that there is always an Umwelt, and a surrounding world of other agents with whom the living subject interacts with. To understand this, Varela construes the hypothesis of "reciprocal specification" (Varela et al. 1991). This is partly inspired by research in psychological categorization (Rosch 1978). On the other hand, it is to some extent similar to the position held by the most prominent representative of embodied phenomenology, Merleau-Ponty (1966).

By stressing the common genesis of phenomenal world and subjective standpoint in perception, Varela tries to overcome the danger of solipsism which every strong constructivist, and to some extent also a biosemiotic position is liable to be subject to (see above). Against this danger, the mutual creation of world and living agent provides the key for understanding why organisms so remarkably fit into their environment, and why communication within it is possible. Here I will limit myself to draw a brief outline of Varela's position and then show, in discussing Merleau-Ponty, some implications which lead, in a backward loop, directly into a semiotic understanding.

On an embodied level, the process of perception is not so different from the fundamental self-limitation of organism, which generates non-self from the encounter with the exterior that has to be excluded from the autopoietic process. This is also valid for perception: The exterior becomes a decisive component in the construction of the percept, and, *vice versa*, the perceiving structure marks decisively the creation of the perceived world. What results is a communion of perceiver and perceived. Here, the presence of the other is reality, not fiction, even though only the process of perception is responsible for its ultimate shape.

From the enactive standpoint, the external world's influence is even augmented by a kind of absoluteness of the objective outside: The real presence of a material perturbation is necessary to engender any perception. Varela et al. (1991 [1995]) try to explain this by analysing the human colour system that seems to be marked by culture *and* by neural structures alike:

> Contrary to the objectivist view, color categories are experiential; contrary to the subjectivist view, color categories belong to our shared biological and cultural world. Thus color as a case study enables us to appreciate the obvious point that chicken and egg, world and perceiver, specify each other. (Varela et al. 1991:172)

This stance is already prepared in Merleau-Ponty. His philosophy is to some extent a semiotic foundation of phenomenology, because the *living body* is arbiter about how the world is perceived (see e.g. Merleau-Ponty 1966:478). This narrow relationship with the biological bases of perception has even lead to the assumption that Merleau-Ponty's work is no phenomenology at all (cf. Latour 1995). Phenomenology, as Husserl construed it, had a strong mentalist tendency in its attempt to recreate the world from pure intentionality. The discipline has nonetheless sparked a field of thought that can be classified as "biological" or "embodiment-" philosophy. From a biosemiotic standpoint, I believe, we can reformulate its general position, where Merleau-Ponty can be added, into the following idea: The living body in its existential concern is the interpretant of the Umwelt's signs. As Merleau-Ponty observes:

An object hence is not really *given* in perception, but experienced and internally reconstructed insofar as it belongs to a world, whose basal structures we find in ourselves and of which it represents only one of the possible concretions. (Merleau-Ponty 1966: 377).

In this view we can see, that an interrelation of phenomenal worlds, which are all bound to the common constraints of biological cognition, or semioses, show the way to a general *conditio vitae* living beings are unified in (Weber 2001a, 2002b). Beyond the fact of supplying the universal framework for biological sense making, the "event space" of the *conditio vitae* is a materially, historically, and semiotically multidimensional conglomerate where signs ever spark new signs, and which is formed by a more ore less intense interdependency of its agents.

This idea, although in a markedly different accent, reminds of the concepts of "hybrids", coined by Latour (1995), a term unifying material entities, cultural customs, and living beings to explain the true characters of the objects of science. It is most interesting, that also Varela has seen this relationship, and even, together with psychoanalyst Amy Cohen, devoted a paper to it (Cohen & Varela 2000). This axis of understanding cannot be underestimated. Latour has observed that scientific constructs are never free of a social bias, but he equally emphasises that they are always full of residues of real bodies. If we consider the hybrid status of Varela's enactivist account (which we have tried to interpret in a biosemiotic fashion), we might have found in his biosemiotic philosophy of cognition, another answer why human world making, be it scientific or explicitly pre-modern, is always an intricate network of bodies and significations. Varela might have supplied the biological grounds to understand our worlds' "surplus of meaning".

9. Shifting paradigms: Biosemiotics as a new episteme for systems biology

Also scientific revolutions might devour their children. In a sense, this seems to be the current scenario in actual genetic research. On the one hand, it is ultimately successful by providing data from genetic sequencing and an ever refined description of genetic regulation (in spring 2001 the coarse sequence of the human genome was published, Baltimore, 2001). On the other hand, modern biology *by this succes* is forced to admit more and more exceptions to the still central dogma of DNA-plasma causal linearity.

This erosion of the very foundations of nucleus-related genetics is due to its own success. Maternal factor, splicing, secondary structure folding, epigenetic regulation (a gene product is feeding back to the genetic apparatus, Keller 2001) are circular rather than linear models of regulation. They display a strong influence of complex plasma reactions, which can change the fate of genetic information (in this sense, already the Jacob-Monod-model showed a first departure from Crick's ideal hypothesis of genetic expression; see also Gilbert & Sarkar, 2000).

Hence, any major medical breakthroughs, which are thought to come along with a broad sequencing of the genome, probably cannot take place before a basic understanding of the post-genomic-regulations will be achieved. Proteomics seem to become the next field of the game, being one to two orders of magnitude more complicated, though (Weber 2001b, Richard Vilems, pers. communication). Most interestingly, this development seems to bring an approximation of genetics and developmental biology, which traditionally is the biological discipline that is concerned with regulation. This is a historically most intriguing outcome: For five decades, advanced genetics seemed to have won over developmental biology in discovering the comparably simple Watson-Crick-model. This advantage apparently is vanishing. In this situation we are forced to reconsider the prophecy of the US Long Range Planning Committee for scientific organisation. For Biology on the long run it expected a paradigm shift from linear interactions to the systems' or "organismic domain" (Strohman 1997, Long Range Planning Committee 1990).

As Emmeche and Hoffmeyer (1991) and before them Blumenberg (1981) have seen already some time ago, neo-Darwinian biology is at odds with giving up a speech that is frankly contradictory for its theoretical framework. By talking of code and message, it contaminates its own theoretical agenda with semiotic terms (a critique which is equally valid for the ever-present teleological arguing in Neodarwinism—the two possibly being strongly linked, Weber 2001, Weber & Varela 2002, and see above).

In this context, it is most interesting that also the modern synthesis of Evolutionary Theory speaks of certain autonomy of the living. Ernst Mayr (1988) even uses the term "vitalism": For him a distinguishing character of organism does exist because the latter is constructed of genetic information. This is certainly a challenge for the autopoietic model, which works without taking genes much into account. Varela stresses autonomy explicitly without recurring to the metaphor of autonomous genetic information. For him self-regulation is primordial. But is not the enigmatic autonomy provided exactly by the genetic program that directs the organism's metabolism (as Schrödinger, though somewhat contradictory, has proposed 1944?) Can organic autonomy be equalled with genetic instruction? Thus, do the genes have an ontological priority?

As is visible in the new problems that arise for genetics, such a view far to easily overlooks the importance of the living body for all biological functioning. The genes, although obviously an abstract code, do not exist in an abstract manner, but are embedded in the functioning of the organism. They are part of this vast circularity: e.g., they are repaired by somatic components, which have been encoded by the units they are repairing. There is no exception of the law that in an organism everything is reciprocally working as medium and final goal. Here, the biosemiotic concept of "genetic scaffolding" provided by Hoffmeyer and his broader view of biosemiotic code duality (Hoffmeyer 1997), supply a framework for solving contradictions.

The crucial point is to have the right philosophy of organism at hand to overcome the intricacies of genetics. In the new shift to systems biology we can not only witness closeness to developmental biology, we also can see a rediscovery of a certain angle of view which is focusing on the minimal living organism as the ultimate object of a *bio*-logic. For a long time this has been the main issue of the organicist school of biology, the paradigm, which was competing with Neo-Darwinism. Now, in the guise of complexity biology as represented by Kauffman (1996), Salthe (1993), Weber & Depew (1996), and Varela, it might offer new competing solutions to the overwhelming challenges of systems biology.

Varela's work might serve as a possible bridge between theoretical biology, organicism, phenomenology and biosemiotics proper. His influential position in cognitive science might facilitate the translation and homogenisation of semiotic concepts, which could lead to a closer acceptance, or "naturalization" of the semiotic argument. In this respect, Varela's "imaginary dimension", has to be considered as a basic aspect of the realization of the living, and at the same time, as a door opened to the human sciences. Varela's work is deeply engaged in cutting down the separation between the two realms, between the human sphere and the remainder of the material world. Based on his work, by exploiting the potential of existential value-genesis as well as the expressive, or performative, aspect of every living being's uttering, a comprehensive cultural-biological semiotics might not be impossible (Weber 2001, 2002b)[4].

10. References

Baltimore, D. 2001. "Our genome unveiled". *Nature* 409: 815.

Barbaras, Renaud 1999. *Le désir et la distance*. Paris : Vrin.

Blumenberg, H 1981. *Die Lesbarkeit der Welt*. Frankfurt am Main: Suhrkamp.

Brooks, Rodney 2001. "The relationship between matter and life". *Nature* 409: 409-411.

Chernyak, L.; Tauber, A.I. 1991. "The dialectical self: Immunology's contribution". In: Tauber, A.I. ed., *Organism and the origins of self*. Dordrecht: Kluwer, S. ?????

Cohen, Amy E.; Varela, Francisco J. (2000): "Facing up to the embarrassment: The practice of subjectivity in neuroscientific and psychoanalytical experience". *Journal of European Psychoanalysis* 10-11.

Deely, J. 1990. *Basics of semiotics*. Bloomington & Indianapolis: Indiana Univ. Press.

Eco, Umberto 1976. *A theory of semiotics*. Blomington & London: Indiana Univ. Press.

Emmeche, C.; Hoffmeyer, J. 1991. "From language to nature: The semiotic metaphor in biology". *Semiotica* 84(1/2): 1-42.

Fleck, L. 1979 [1935]. *Genesis and development of a scientific fact*. Chicago and London: Univ. of Chicago Press.

Gilbert, Scott F.; Sarkar, S. 2000. "Embracing complexity: organicism for the 21st century". *Developmental Dynamics* 219: 1-9.

Goodwin, Brian 1997. *Der Leopard, der seine Flecken verliert. Evolution und Komplexität*. München: Piper.

Hoffmeyer, Jesper 1997. *Signs of Meaning in the Universe*. Bloomington: Indiana University Press.

Jonas, Hans 1973. *Organismus und Freiheit. Ansätze zu einer philosophischen Biologie*. Göttingen: Vandenhoeck und Ruprecht.

Kauffman, Stuart 1996. *At home in the universe: The search for laws of self-organization and complexity*. London: Penguin.

Keller, Evelyn Fox 2001. *Das Jahrhundert des Gens*. Heidelberg: Campus.

[4] Research on which this paper is based has been kindly supported by scholarships from the French Government and the *Deutsche Bundesstiftung Umwelt*.

Kull, Kalevi 2000. "An introduction to phytosemiotics: Semiotic botany and vegetative sign systems". *Sign Systems Studies* 28: 326-350.

Lakoff, John; Johnson, Mark 1999. *Philosophy in the flesh*. New York: Basic Books.

Latour, Bruno 1995. *Wir sind nie modern gewesen*. Frankfurt am Main: Fischer.

Lenoir, Timothy 1982. *The strategy of life: teleology and mechanics in 19. century German biology*. Studies in the history of modern science, Vol. 13. Dordrecht: Reidel

Long Range Planning Committee 1990. "The next revolution in biology will be in the integrative or organismic domain". In: "What's Past is Prologue. A White Paper on the Future of Physiology and the Role of the American Physiological Society in It". *The Pysiologist* **33**:161-180.

Löwy, I. 1991. "The immunological construction of the self". In: Tauber, A.I. ed., *Organism and the origins of self*. Dordrecht: Kluwer, S. 43-75.

Margulis, L.; Guerrero, R. 1991. "Two plus three equal one. Individuals emerge from bacterial communities". In: Thompson, W. I., ed.: *Gaia 2. Emergence. The new science of becoming*. Hudson, NY: Lindisfarne Press.

Maturana, H. 1987. "Kognition". In Schmidt, S.J., ed.: *Der Diskurs des radikalen Konstruktivismus*. Frankfurt am Main: Suhrkamp, S. 89-118.

Maturana, H. R. & Varela, F. J. 1980. *Autopoiesis and cognition: The realization of the living*. Boston: D. Reidel. German translation 1987: *Der Baum der Erkenntnis. Die biologischen Wurzeln menschlichen Erkennens*. München: Goldmann.

Mayr, Ernst 1988. *Toward a new philosophy of biology*. Cambridge, Mass: Harvard Univ. Press.

Merleau-Ponty, Maurice 1966. Phänomenologie der Wahrnehmung. Göttingen: de Gruyter.

Prigogine, Ilya.; Stengers, Isabelle. (1990): *Dialog mit der Natur*. München: Piper.

Rosch, Eleanor 1978. "Principles of categorization". In: Rosch, E.; Lloyd, B.B. eds., *Cognition and categorization*. New Jersey: Hillsdale.

Rose, Steven P.R. 1998. *Lifelines. Biology beyond determinism*. Oxford: Oxford University Press.

Salthe, Stanley N. 1993. *Development and Evolution: Complexity and Change in Biology*. Cambridge, Mass: MIT Press.

Schrödinger, Ernst 1991 [1944]. *What is Life? The physical aspect of the living cell*. Cambridge: Cambridge Univ. Press.

Sebeok, Thomas A. 1976. *The sign and its masters*. Lanham & London: Univ. Press of America.

Steiner, G. 2001. *Grammatik der Schöpfung*. München: Hanser.

Stjernfelt, Frederik 1992. "Categorial Perception as a General Prerequisite to the Formation of Signs?" In: Sebeok, T.A.; Umiker-Sebeok, J. (eds.): *Biosemiotics: The Semiotic Web*. Berlin and New York: Mouton de Gruyter.

Strohmann, Richard 1997. "The coming Kuhnian revolution in biology". *Nature Biotechnology* 15: 194-199

Varela, Francisco J. 1991. "Organism: A meshwork of Selfless Selves". In: Tauber, A.I. ed.; *Organism and the origins of self*. Dordrecht: Kluwer.

— 1996. The early days of autopoiesis: Heinz von Foerster and Chile, *Systems Research* 13: 407-417

— 1997. "Patterns of Life: Intertwining Identity and Cognition." *Brain and Cognition* 34:72-87.

Varela, F. J., Anspach, M. 1991. "Immuknowledge: the process of somatic individuation", in: Thompson, W. I., ed.: *Gaia 2. Emergence, the new science of becoming*. Hudson, NY: Lindisfarne, S. 68-85.

Varela, F. J.; Thompson, E; Rosch, E. 1991. *The embodied mind. Cognitive science and human experience*. Cambridge, Mass.: MIT Press.

Weber, Andreas 2001a. "Cognition as expression. The autopoietic foundations of an aesthetic theory of nature". *Sign System Studies* 29.1:153-168.

— 2001b. Das verkaufte Volk. Magazin der Süddeutschen Zeitung, 27.11.

— 2002a. "Feeling the signs. Organic experience, intrinsic teleology and the origins of meaning in the biological philosophy of Hans Jonas and Susanne K. Langer". *Sign System Studies* 30.1: 183-200.

— 2002b. *Natur als Bedeutung. Versuch einer semiotischen Ästhetik des Lebendigen*. Dissertation, Humboldt-Universität zu Berlin.

— in print: "Turning the inside out. Natural forms as expression of intentionality". *Proceedings of the Finnish Semiotic Society*.

Weber, A. & Varela, F.J. 2002. "Life after Kant. Natural purposes and the autopoietic foundations of biological individuality." *Phenomenology & the Cognitive Sciences*, forthcoming.

Weber, B.H.; Depew, D.J. 1996. "Natural selection and self-organization. Dynamical models as clues to a new evolutionary synthesis". *Biology and Philosophy* 11: 33-65.

accepting grief

we sit in silence
in the space of
his lost presence
one lone candle
bright in the daylight

a drop of wax
wells and brims
flowing and congealing
as textured candle
to melt again
fueling flame
as the candle burns

a warm tear
wells and brims
grief flowing and congealing
as textured being

grief, a new ground
for the living flame
illuminating living

Cybernetics & Human Knowing, Vol.9, No.2, 2002, pp. 31–47

Varela and the Uroborus:
The psychological significance
of reentry

Terry Marks-Tarlow[1], Robin Robertson[2], and Allan Combs[3]

Abstract: The Uroboros, or snake that swallows its own tail, symbolizes regeneration and renewal the world over. It was adopted by Francisco Varela as an icon for his re-entry term in "A Calculus for Self-Reference." The present paper examines how the notion of re-entry can be applied psychologically, to issues of autonomy and identity. According to Varela, all autonomous systems are structurally open and functionally closed, which leads to paradoxical qualities particularly evident in higher order cybernetics. First Varela's early work is placed within a philosophical and historical context. Next, notions about biological autonomy are examined from the perspective of nonlinear dynamics. Then, the recursive dynamics of consciousness are explored through social mirror theories of identity formation. Finally, Varela's ideas are applied self-referentially to descriptions of his own experience as he neared death.

"We are led to consider in all seriousness the traditional image of the snake eating its own tail as the guiding image for autonomy as self-law and self-regulation." (Francisco Varela, 1979)

This paper was written in tribute to the work of Francisco Varela. In the spirit of the Uroboros it explores Varela's concept of reentry in a number of different contexts, all of which make possible fluid but stable self-identity in systems that are structurally open yet functionally closed.

In myth the world over, the Uroboros, the snake that swallows its own tail, symbolizes self-generation and renewal. In ancient Egypt, the Uroboros is depicted on tombs as guardian of the Underworld. There it represents the liminal moment when death encounters resurrection. In the West African vodun religion of Benin, the Uroboros appears as an icon for the god of order, Dan. Dan mediates between the world of the gods and our world, in order to create order in the wind, water and other cyclic, often chaotic, rhythms of nature. In old European maps the Uroboros swims in distant seas to symbolize the fearsome edges of the known.

Over time the symbol of the Uroboros has evolved from the realm of matter to that of spirit. This symbol follows a similar trajectory to the mathematical and scientific research of the late Francisco Varela, as it evolved over the course of his lifetime. Recognizing the deep significance of the Uroboros to processes of

[1] Private Practice, Santa Monica, California, USA. Email: markstarlow@hotmail.com

[2] General Editor, *Psychological Perspectives*. Email: rrobertson@pacbell.net

[3] University of North Carolina—Asheville and Saybrook Graduate School—San Francisco, USA. Email: combs@unca.edu

self-generation, Varela chose an icon of the snake swallowing its own tail as a symbol for reentry in his calculus of self-reference. It depicts what Varela describes as the "ceaseless circular process which is, in fact, the symbol which tradition has chosen to represent the creation of everything since time *immemorial*" (Varela, 1975, p. 23).

Abstract Origins: A Calculus for Self-Reference

Spencer-Brown's vision, then, amounts to a subversion of the traditional understanding on the basis of descriptions. It views descriptions as based on a primitive act (rather than a logical value or form), and it views this act as being the most simple yet inevitable one that can be performed. Thus it is a non-dualistic attempt to set foundations for mathematics and descriptions in general, in the sense that subject and object are interlocked. From this basic intuition, he builds an explicit representation and a calculus for distinctions. (Varela, 1979, p. 110).

In a seminal early paper, "A Calculus For Self-Reference," (1975), Francisco Varela extended G. Spencer Brown's *Laws of Form* (1969); he further extended this work in *Principles of Biological Autonomy* (1979). In the former, Varela said that: "By succeeding in going deeper than truth, to indication and the laws of its form, he [Spencer-Brown] has provided an account of the common ground in which both logic and the structure of any universe are cradled" (Varela, 1975, p. 6).

Spencer-Brown's work was critical for Varela in two significant ways. First, *Laws of Form* begins with the realization that *indication* and *distinction* are inextricably entwined: when any indication is made, any mark, it automatically creates a distinction between two states: that which is marked and that which is not. This is central in Varela's concept of autopoiesis. Varela (Varela, 1979) expresses the unity of distinction and indication this way:

A *distinction* splits the world into two parts, 'that' and 'this,' or 'environment' and 'system,' or 'us' and 'them,' etc. One of the most fundamental of all human activities is the making of distinctions. (p. 84)

But,

A distinction cannot exist without its concomitant value. The distinction thus becomes an *indication*, i.e., an indication is a distinction that is of value. (p. 84, 107).

Taking these two states–the marked and not-marked–together with two seemingly unarguable "laws" (i.e., primitive axioms), Spencer-Brown developed a *calculus of indications* which he then extended into an algebra. If that algebra, which is fully general for any 2-valued system, is assumed to deal with logic, it is identical to Boolean algebra. But his calculus is more basic than logic.

The second significance of *Laws of Form* for Varela was that, beginning with the simplest possible system of mathematics, which depended only on an act of indication and distinction, Spencer-Brown eventually arrived at self-reference. In solving actual problems involved with the routing of trains, Spencer-Brown found that he was forced to develop algebraic equations within his calculus in which the function referred to itself. This sometimes led to equations whose resolution was first the marked state, then the unmarked stated, ad infinitum.

After much thought, he came to realize that this situation was equivalent to imaginary numbers in normal mathematics; that is, numbers which arose out of normal equations, but were something other than numbers as we normally know them. In mathematics, imaginary numbers eventually came to be interpreted graphically as numbers that existed at a right angle to the normal number line; hence not on the number line itself. Similarly, Spencer-Brown came to interpret his strange re-entrant numbers as oscillations in and out of the space created by the marked stated and the unmarked state. Thus he felt that this marked the creation of *time* within the space created by his calculus. Varela called this insight "one of his most outstanding contributions" (1979, p. 138).

As a mathematician and logician, "Spencer-Brown wisely finished his work at the point when self-reference entered the picture, satisfied with the deep insight that self-reference introduces time" (Robertson, 1999, p. 53). Varela, a biologist and cybernetician, decided to begin where Spencer-Brown ended by adding self-reference as a third primitive value in a 3-valued calculus. He was proposing that, in the world we actually live in, there is not only the act of indication, which distinguishes between two states, there is also an act that endlessly reaffirms identity. By proposing that form re-enters itself at the most basic level, Varela identified a deep principle he believed went beneath logic itself to the very creation of structure in the universe.

From Duality to Trinity, then Re-Entry

Consider the following two statements:

> When Robert Frost was asked about the organization of a collection of poems, he is supposed to have replied that if a book consisted of 29 poems, then the book itself had to be the 30th poem. A collection, in other words, must transmit artistic integrity as eloquently as the individual works within it. No grab bag of hits will do (Hampl, 1999).

> For every (Hegelian) pair of the form A/not-S there exists a star where the apparent opposites are components of the right-hand side (Varela, 1979, p. 101).

It might not be obvious that both quotations are saying the same thing, but Varela realized that supposed opposites are all created through a Spencer-Brown distinction and thus are inextricably joined through a third, the indicative act that creates them. There is always a whole on a meta-level above the opposites which binds them into a greater whole. It matters little whether we are talking about dividing a whole into two parts or into 29 parts, like Frost's collection of poems, because once you have this realization, you inevitably have the possibility of self-reference. Any one of Frost's poems, for example, can also refer to the collection in which it is contained.

The importance of the third, the act that creates and binds the components, is also familiar to the readers of these pages in the work of the American logician Charles Peirce (e.g., 1998), who built a three-factor logic surprisingly compatible with Spencer-Brown's two-value system. Peirce's logic was also the basis of a major theory of semiosis, and will be useful later for discussing self-reflective consciousness. For these reasons we will briefly introduce it here for those who are not already familiar with its basic structure. Peirce's logic involves three parts, in his terms a *first*, a *second*, and a *third*. In general, for any argument the beginning is the first, the end is the second, and the middle term that expresses the relationship of the first to the second is the third. For instance, the possibility of seeing a bluebird while walking in the park is first, the bluebirds actually seen are second, and the probability functions that relate the first to the second are third. In logic as in language the third expresses a general or abstract relation of the first to the second. It is the medium or connection that binds them. In semiotics, the third is the concept that carries the meaning of this relationship. For example, the warm and slightly bitter, yet pleasing taste in my mouth (first or the representamen/primary sign), turns out to be tea (second or object), and I reflect on the fact that it is a black Indian variety (third or the interpretant). The third is also a *sign*. It is a more developed idea of the taste than first now it is for instance categorized as a certain type of tea. On reflection it is apparent that each interpretant can itself be a representamen for other interpetants. One might consider the whole range of black teas that are available in a good café, or the all poems in a book, as in the above example and the choice of te a person makes when tasting several is a sign of his taste. In these instances we see that signs convey *meaning* and the role that meaning plays in language, logic, and consciousness. It is easy to imagine networks of signs that reflect back onto each other in complex re-entrant loops, an idea closely related to self-reflection in language and consciousness, and to which we will return below.

Varela was interested in the notion of re-entry at many levels of living systems, ranging from cell metabolism, to cognitive processes in thoughts, words, and self-aware consciousness. In all of these self-reference is central to the deepest levels of organization. Despite the awkwardness of "the brain writing its own theory, a cell computing its own computer, the observer in the observed" (Varela, 1975, p. 5), re-entry of any such system back into itself is what allows each to operate autonomously. By being self-indicative within a given domain, whether at biological, cognitive, or experiential levels, these systems become capable of maintaining their identity separate and distinct from their environment without outside intervention. In Varela's words:

> . . . closure and the system's identity are interlocked, in such a way that it is a necessary conse-
> quence for an organizationally closed system to subordinate all changes to the maintenance of

its identity" (1979, p. 58). And it can only do that through self-referential feedback of its identity to itself. Cybernetics thus evolves naturally into second-order cybernetics.

Varela originally joined forces with his mentor Humberto Maturana (Varela, Maturana & Uribe, 1974), to develop Maturana's concept of autopoiesis, "in cybernetics, a term coined by Humberto Maturana for a special case of homeostasis in which the critical variable of the system that is held constant is that system's own organization (sic)" (Bullock, Stallybrass & Trombley, 1999). Varela further articulated these initial concepts in his *Principles of Biological Autonomy* (1979). Its major thesis is that biological systems become autonomous, capable of self-generation, by remaining *structurally open yet functionally closed*. The central idea that makes this possible is that a system can re-enter itself endlessly, i.e., create repetitive loops in which end products are fed back into the system as new starting points.

Here Varela is presenting a new answer to the age-old question of how form emerges in a formless world. This is one of the great issues in the history of Western philosophy, and worth a brief regression.

How does Form Emerge? A Digression into the History of Ideas

You cannot step twice into the same river, for fresh waters are ever flowing in upon you.
— Heraclitus (Copleston, 1962, p. 55).

From time immemorial, men and women have noticed that the world is in flux, constantly changing. The most famous expression of that verity was Heraclitus' observation. Yet it is also true that, despite the river constantly flowing, it remains a river, separate from the land through which it flows. How can these two be reconciled? How can form exist in an ever-changing world?

The two classic answers to the question of how form emerges in nature are those of Plato and Aristotle. This shortest summary of their positions (which, of course, oversimplifies both) is that Plato was concerned with an eternal world of *being*, which undergirds our changeable physical reality, Aristotle with the world of becoming, in which we live. For Plato, all that exists in physical reality is a pale imitation of eternal forms, ideas, which are unchanging. For Aristotle, what we see in nature is matter in the process of developing into its predestined form. If that sounds too much like Plato's position, it's because Aristotle's position is, in fact, less clear than Plato's and is informed largely by his studies of biology. There he saw oaks growing from acorns, chickens from eggs, and so on. It seemed clear to him that the ultimate form of the oak had to already be contained within the acorn, and the chicken within the egg. In essence, his was a teleological position, Plato's a causal one.

Later Christian philosophers struggled to achieve a satisfactory resolution of these two positions. During much of this period there was little or no interest in looking at the world itself to see what was actually there. With the dawn of the

Renaissance, however, the focus of art and literature shifted to the real world, and the type of detailed observation that would lead to the birth of modern science came into existence. As the latter emerged in the seventeenth century, careful observations of nature and detailed records of these observations slowly began to erode the faith in the old assumptions about nature espoused by Plato and Aristotle. Meanwhile, a growing emphasis on objective observation began to separate the observer from the observed, while at the same time an increasing number of individuals began to engage in more or less objective self-observation as well (Combs, 2001). This ability for reflection already reached a high point in René Descartes, who practically invented the notion of consciousness in his mid 17th century *Meditations*. In the early 18th century Bishop George Berkeley carried Descartes' introspective orientation to the limit, arguing for the *idealist* view that all sensory experiences are presentations of the mind. This line of thinking was not Platonic by any means, but did not contradict it either. On the other hand, the more successful stream of British empiricism represented by John Lock's notion of *tabula rasa* had a distinct Aristotelian flavor, with its emphasis on the primacy of the material world. Later in the century Immanuel Kant made a real advance on the ideas of Plato and Aristotle.

Kant argued that Locke was right, there is a physical world and it is composed of physical entities, *das ding an sich* (the thing in itself). But he also agreed with Berkeley that all we can experience are the ideas in our mind. But Kant saw that this was a psychological issue. We don't accumulate a direct portrait of reality from our sensory experiences as Locke thought, we construct reality using in-born categories of mind. Those categories, like Plato's ideas, might (or might not) be eternal, but reality isn't simply a poor mirror of those eternal forms. Those ideas (*categories* for Kant) are in the mind, and are what we use to structure our experience of reality. With the considerable success of this notion Kant instigated a kind of Copernican Revolution in philosophy according to which what we can know is never the thing in itself, but always the thing as represented. Never a perfect copy, it is always the outcome of our engagement with the external world and the active power of the mind. Our knowledge of the world bears the stamp of our own conceptual activity (Putnam, 1990, p. 261).

While influential among philosophers, especially on the Continent, eighteenth century science took little notice of Kant. It was too early for science to worry about issues like the relationship between mind and world. Indeed, it is only recently that science has had to address such issues through detailed study of the brain and its relationship to the world.

All the above issues come to focus in the work of brain scientist and philosopher Walter Freeman, whose research on the brain provides exquisite support for Varela's model of biological systems as structurally open and functionally closed, reentering themselves endlessly to preserve identity. Freeman and his students studied the olfactory system in rabbits, attempting to trace every step of neural activity from the initial sensory receptors through all the paths this activity took until it reached the cerebral cortex. At that point, suddenly the original neural activity vanished! In Freeman's words: "just like the rabbit down the rabbit hole in *Alice in Wonderland*." Instead, they found a totally "new pattern of cortical activity that was created by the rabbit brain." They were able to replicate this finding, not only in the olfactory system, but in all the other major senses as well. No matter which sense they studied, when the neural activity reached the cortex, suddenly something new appeared, and that new pattern of activity varied each time they repeated the experiments. Freeman concluded that "the only knowledge that the rabbit could have of the world outside itself was what it had made in its own brain" (all quotes from Freeman, 1995, p. 2).

For a rabbit brain, and by extension, the human brain, the most complex biological structure on the planet, Freeman appears to have amply demonstrated Varela's conjecture that biological systems become autonomous, capable of self-generation, by remaining *structurally open yet functionally closed*. How else to describe Freeman's rabbit, whose brain receives sensory input from the world outside, yet responds with a creative response from within? Freeman explains this in terms of intentionality. He views the brain as an intentional structure that reaches out into the world through the senses, rather than merely passively accepting sensory input. "I explain the lack of invariance [of our experience of the world] as based on the unity of intentionality, such that every perception is influenced by all that has gone before" (Freeman, 1995, p. 2). Curiously, Freeman himself is critical of Varela's position, which he interprets as "metaphysical solipsism, according to which everything that exists is the projection of a brain" (Freeman, 1995, p. 2). But though he disagrees with Varela's view, it is difficult not to see them both, each in their own way, as in agreement with Kant that all we can know is "a joint product of our interaction with the external world and the active power of the mind" (Putnam, 1990, p. 261).

The Soft Boundaries of the Self

The boundaries of my body are invisible, a floating shield of self-production, unaware of space, concerned only with permanent bonding and unbonding. The self is also an ongoing process every time new feed is ingested, new air is breathed in, or the tissues change with growth and age (Varela, 2001, p. 263)

Varela's concept of biological identity contains paradoxical elements, implying that a system is open precisely because it is closed and closed because it is open. That is, biological systems retain cohesive identity because, as Prigogine and Stengers (1992) might say: they exist in far from equilibrium conditions with an exchange of matter, energy and information across open boundaries. At the same time, this identity can evolve dynamically precisely because of the system's autonomous functioning. Hardly the metaphysical solipsism of which Freeman accuses him.

As his career progressed, Varela extended his work from biological into psychological realms. During the past two decades, Varela helped to spearhead a shift within cognitive science away from the classical view, resembling Locke's position, of an inner mind representing an outer world using symbols in a computational language of thought. Instead, Varela embraced what he and colleagues called an *enactive* view of cognition (see Varela, Thompson & Rosch, 1991). This approach blends concepts from phenomenology and non-linear dynamics. As most recently conceptualized by Thompson and Varela in a forthcoming book, enactive cognition involves the following three elements:

1. *Embodiment*: The human mind is not confined within the head, but extends throughout and even beyond the living body to encompass the world outside of the organism's physiological boundaries.
2. *Emergence*: Human cognition emerges through self-organized processes that span and interconnect the brain, body and environment in reciprocal loops of causation. In addition to the 'upwards' causation of personal consciousness by neural and somatic activity, there is the 'downwards' causation of neural and somatic activity by the person as an active, conscious agent.
3. *Self-Other Co-Determination*: Because open boundaries exist at all levels, which include the social, the individual human mind does not emerge in isolation, but instead is embedded within an interpersonal context. Through ongoing, dynamic interaction, self and other create one another at the most fundamental levels.

This view of cognitive autonomous functioning places the body, physical environs and even the interpersonal environment all within the purview of a kind of extended subjectivity. Now we can see how Varela extends the notion of "structural openness" from biology into psychology. At the psychological level, autopoiesis involves the evolution and dynamics of a psyche that employs structural openness, i.e., growth, change, movement in time, plus social and environmental coupling, in order to achieve functional closedness, i.e., stable, cohesive integrity and identity across time. The arc of consciousness moves outside and away from the body into the world and then back inside again, in the form of perceptual-motor loops and an extended phenomenal presence. Freeman, in a lovely expression of this social embeddedness of mind likens the interchange between two brain/minds to a dance. "Dancing with others is the quintessential means to bridge the solipsistic gulf" (Freeman, 1995, p. 5). We see this throughout the normal experiences of life. When two strangers interact, it's easy to view them as "functionally closed," with communication limited to the content of the words they speak. Contrast that with lovers whose whole world is the other person. Looked at from without, it's easy to see the "dance of love" they go through.

This mutual dance is more than anecdotal. In his *Beyond Culture* (1977), linguistic anthropologist Edward T. Hall presented research on *synchrony* by experimental psychologists William Condon and Ray Birdwhistell. They filmed couples and larger groups in a wide variety of generalized behavioral settings. When, afterwards, they slowed down the films, what looked in normal time like spontaneous behavior, was actually a dance in which each gave and responded to tiny behavioral cues. According to Hall, "Viewing movies in very slow motion, looking for synchrony, one realizes that what we know as dance is really a slowed-down, stylized version of what human beings do whenever they interact" (Hall, 1977, p. 72). He concludes further that this ability to respond to tiny behavioral cues is innate and well established by the second day of life, if not in fact present from birth.

More recent infant research suggests that this synchrony is involved in the development of self by the infant, and in the mother's development as well (Beebe & Lachmann, 1988; Beebe, Jaffe, & Lachmann, 1992). Here, the expanding selves of child and mother are integrally linked. How the child comes to view him- or herself depends critically upon mother's emotional responsiveness. How the mother views herself depends critically on that of the baby. One is reminded of what Freeman calls a "Society of Brains." In this dance of synchrony the infant and mother become a society of two, in which fundamental perceptual and motor choices represent the elemental distinctions, or signs, that in developmental time will lead to the higher order and more abstract signs we associate with meaning and consciousness. This escalating web of signs will eventually lead to the ability to become an "objective" observer, as we will see below.

Returning momentarily to infancy, however, let us note that as conscious, its hallmark will be the capacity to draw distinctions between one thing and the next. And, in order to draw distinctions, opposites must separate in consciousness, allowing the child to experience each case independently in time. Unlike the realm of the unconscious, opposites do not necessarily co-exist comfortably in the conscious mind. For with the capacity to separate opposites in consciousness, also comes the possibility for contradiction. No longer do opposites simply merge and emerge from one another; now they may actually collide.

Yet as Varela (1975) intuited, self-reference enfolds and sometimes incorporates opposite states. That is, the recursive enfolding of opposite states leads to a system that is functionally closed, but not necessarily internally contradictory, even at the cognitive level. Contradiction can be avoided when opposite states continue to be expressed sequentially in time as oscillating emotional and/or experiential states, or when they are separated by barriers between conscious and unconscious levels of awareness. Isn't this reminiscent of Spencer-Brown's oscillation in and out of the form?

At the cognitive level, recursive dynamics lead us ineluctably to second order cybernetics. Consciousness operates continually upon objects and aspects of both internal and external worlds. Reentry occurs through feedback processes whereby products of one state of consciousness become the content for the next state. Each sign becomes the *object* of another sign, and these contain each other like a reflection captured between the surface of a pond and a mirror held above it. As one of the authors has previously noted, in this moment consciousness crosses a threshold of complexification and self-awareness is born (Marks-Tarlow, 1999). In poet John Donne's words: "The beast does but know, but the man knows that he knows."

Self and Recursion in Social Theory

There exist a number of social theories that contain recursive dynamics and identity formation at their cores. In each, consciousness arcs away from the self towards the other, only to return again through social interaction. Out of these recursive loops of social interaction a global sense of self is shaped and held in a coherent self-identity. Object relations ego theorist Jane Loevinger (1976) calls such processes "circular reactions." Similar notions were expressed a century ago by Charles Horton Cooley (1902/1968), who used the phrase "looking-glass self" to describe how a sense of oneself is integrally connected to a correlate sense of others. I see you looking at me, and observing you I begin to sense the me with whom you are interacting.

At about the same time in history the American psychologist James Mark Baldwin proposed the similar idea that infants first learn to discriminate people from other objects in the environment before recognizing themselves as people. Only later, however, does the child finally distinguish in others the same feelings experienced in his or herself, to arrive at full social awareness and empathy. Baldwin's model is similar to Cooley's, but in it the self forms the mirror in which one knows the other, rather than visa versa. The two models appear to be reverse images of each other.

On reflection, the question of whether it is the self that forms a mirror for the other, or the other that forms a mirror for the self, acquires the aspect of a chicken and egg issue, merely demonstrating the reversibility and yet the ultimate equivalence of marked and unmarked states. Each sign is the object of another sign, itself the object of reflection.

Perhaps the most famous example of circular reactions is found in George Herbert Mead's (1934) theory of "social behaviorism." Like Cooley, Mead postulated that mind and consciousness arise as children learn to represent themselves to themselves through taking the views of others such as the mother or a teacher. Later, children acquire an awareness of a *generalized* or collective *other*.

According to Mead, the development of self-reflective awareness involves a sense of 'I' (as an active subject) and a sense of 'me' (as an object acted upon by others). We can see this theory in action by watching young children learning to speak self-referentially. First they refer to themselves by name, in the third person, as if they were objects seen through the eyes of others. Only later do they gain the ability to refer more directly to themselves, as subjects in the first person form of "I."

Similar ideas are found in the writings of psychoanalytic object relations theorists such as Melanie Klein or Margaret Mahler, who postulate that our extended system of early interpersonal relations becomes internalized as aspects or representations of self. Thus, the developing awareness of self and others go hand-in-hand. In a similar, though evolutionary, vein, Francisco Happé (1998) speculates that during the process of human evolution, homo sapiens became conscious as a side effect of selection pressure to read the minds of others. This theory is based in part on Happé's observations of autistic development, where social deficits are seen to accompany deficits in self-awareness. Social mirror theory is also supported by current neurological research that identifies the presence of 'mirror neurons' in the ventral frontal pre-motor area of macaque monkeys (Rizzolatti, Fadiga, Gallesi & Fogassi, 1996). These are neurons that fire in similar patterns when a monkey performs a simple action such as grasping and lifting a cup, or observes another monkey (or even robot arm!) doing so. Mirror neurons have also been recently identified in humans as well (Nishitani and Hari, 2000).

As suggested by the presence of mirror neurons, the development of functionally closed, autonomous identity out of structurally open cycles of interactions does not occur only in early childhood, but extends throughout life, we continually rely on the internalizations of our perspectives of others to retain stable, cohesive identities. As we do so a sense of the subjectivity of the others is co-created along with that of the self, leading to complex webs of intersubjectivity.

In all the above models of circular reactions, self-awareness and identity emerge in a functionally closed, autonomous, fashion from structurally open cycles of social interaction. Internal complexity builds as awareness of oneself develops hand-in-hand with our awareness of others. As cycles of interactions between self and other become recursively enfolded into self-self cycles, a fully symbolic experience of self emerges from more concrete levels of self-reflection. In Peirce's terms, increasingly complex networks of signs are created that reflect each other like jewels in the Net of Indra, and out of which arise the emergent phenomenal reality of the self. In the words of Søren Kierkegaard (1949), written in 1849, the self: "is a relation which relates itself to itself [and in such]...the relation is then the positive third term, and it is the self" (p. 216).

Social mirror theories help us to see how new, more complex and abstract levels emerge from infinite, recursive cycles of consciousness re-entering itself. This process peaks in our own capacity psychologically and linguistically for self-referentiality, the ability to use symbols like the "I" form of speech, to represent oneself as a symbol.

Recursive Loops in the Development of Intelligence and Self

From the above themes it is perhaps not surprising that a growing sense of self-awareness and objectivity is essential to psychological growth, and indeed to the developmental unfolding of intelligence itself. A survey of the work of many developmental psychologists makes it clear that an essential feature of intelligence is the ability to reflexively stand back, as it were, from our own experience and see it objectively (e.g., Cook-Greuter & Miller, 1999; Fischer & Harter, 1999; Kegan, 1982, 1994; Kohlberg, 1981; Loevinger, 1976; Flavell,

1963; Wade, 1996). Much of development can be seen in this light, and the changing sense of self that comes with this increasing objectivity is tantamount to nothing less that a series of transformations of our identity as we mature. This motif can be carried into the realm of consciousness studies where it bears directly on the spiritual development as well (Combs, 2001; Gebser, 1949; Wilber, 1999).

For instance, Piaget's preoperational stage is associated with a period of magical thinking and identity with the physical body. The child at this stage *is* his or her body. During the next stage, concrete operations, identity shifts to become embedded in a social context. We *are* the role we play in the family and society. This is the level of Kohlberg's "good boy, nice girl" morality, and of conformity-oriented moral behavior. The transition to Piaget's formal operations period is linked with a shift of identity to an inner self, and an absolutist morality. These changes go hand in glove with changes in the ability of the individual to stand back and observe his or her own process. While the magical child is identified with his or her own body, the slightly older conformity-oriented child stands back from the body, seeing it objectively as an articulated aspect of his or her experience, and identifies with the group. Further into the formal operations period he or she will see the group objectively and come to identify with an inner self. Theorists of the history of consciousness (Barnes, 2000; Beck & Cowan, 1996; Gebser, 1949; Wilber, 1999) observe that this sequential disentanglement with body, the group, and finally the mental process itself, at the highest level of formal operations and in post-formal-operations intelligence, characterizes the history of human consciousness itself as it moved from hunting and gathering (magical) societies to agricultural one (mythical or role oriented societies), and on to the ancient civilizations such as Greece, Rome, and the Hebrews (formal operations), and after a regression during the Middle Ages, to the modern and post-modern worlds (post-formal operations).

This ability to stand back from the body, from the mind, from increasingly large fields of thought, thereby appraising concepts and attitudes as objects of reflection, and finally from emotions and even our very sense of identity, marks a trajectory that ties together the fields of psychological development, historical evolution, and spiritual unfolding. The religious scholar Frithjof Schuon (1981) observes that,

> There is no knowledge without objectivity of the intelligence; there is no freedom without objectivity of the will; and there is no nobility without objectivity of the soul. (p. 15)

Objectivity is the fruit of self-awareness, and surprisingly enough, self-awareness is the gift of complexity. Reflecting at length on this matter, psychoanalyst Stanley Palombo (1999) concludes that it is through complex networks of interactions, within the psyche and within the brain, that our sense of self comes to have access to the many thoughts, feelings, and motivations that otherwise drift about as disconnected attractors, controlling us like puppets without our knowledge or understanding. In other words, objectivity requires wholeness while the usual human condition involves significant fragmentation. Motivational aspects of the mind are only loosely connected to cognitive belief systems, rational process, perceptions, and even emotions. The idea could also be stated in terms of the interconnections of neural networks in the brain and the attractors that their activities trace. Palombo points out that it is the goal of psychotherapy—and here we would add, the goal of psychological growth in general—to connect these fragments into a more complex, fully integrated, system of processes in which few psychological events continue on their own outside of consciousness.

Years ago the pioneer of biofeedback, Elmer Green (Green & Green, 1977) observed that when a system is supplied with feedback it somehow tends to regulate itself to increasing degrees of balance and stability. For example, biofeedback for heart activity or blood pressure does not usually influence them in an immediately useful way, but over time they become stabilized with greater balance and health. There are many body therapies today, from progressive relaxation to Network Chiropractic, that seem to work on the same principle, namely increasing the patient's awareness of different parts of aspects of their bodies, and thus lending them a natural stability. The situation is not much different in the psychological realm. The 20th century sage and yogan Sri Aurobindo (1972) once observed,

> All developed mental men, those who get beyond the average, have in one way or other, or at least at certain times and for certain purposes, to separate the two parts of the mind, the active part, which is a factory of thoughts and the quiet masterful part which is at once a Witness and a Will, observing them, judging, rejecting, eliminating, accepting, ordering corrections and changes, the Master in the House of Mind. (p. 83)

Modern development theorists tell us much the same thing in less poetic language. Growth comes with increasing complexity, which leads to increasing objectivity. Such objectivity comes with a growing ability to see larger and larger chunks of whatever it is that is us in objective ways via reflective consciousness. In doing so, our own sense of self becomes larger and less constrained. Yet all the while, the increasing objectivity of stepping back from ourselves retains paradoxical elements tying it inexorably with the increasing subjectivity of coming into our selves. For mystics, this growing sense of self, shedding previous identities like the snake skins of Uroboros, leads to larger and larger vistas of experience.

Recursive Interpenetration: Varela's Final Inconclusion

One tries to get inside oneself
that inside of the outside
that one was once inside
once one tries to get oneself inside what
one is outside:

to eat and to be eaten
to have the outside inside and to be
inside the outside.

R. D. Laing (1970, p. 83).

Varela was a Buddhist, a student of the Tibetan master Chögyam Trungpa. This meant that he was very familiar with Buddhist insight meditation, perhaps the most highly developed technique ever devised for cultivation of self-observation. This in mind, it is perhaps not surprising that towards the end of his life he shifted his professional interests from cognition viewed as abstract entity toward the subjective essence of consciousness itself. He embraced the work of phenomenologists such as Edmond Husserl, who systematically studied the structure of experience. In the end, Varela wound up embodying his own ideas and applying the tools of his profession, that began as a biologist, to his own subjectivity.

In his final paper of sole authorship, Varela (2001) explored the phenomenology of his own struggle with a shattered sense of self that accompanied his severe medical condition of a liver transplant and subsequent failure. The paper is entirely self-referential. Varela used his deteriorating physical condition to explore in himself the moving boundaries of an identity that paradoxically sometimes rejected the self as alien and embraced the alien as self. He implicitly highlighted the paradoxical aspect of his condition with the contradictory title, *Intimate Distances: Fragments for a Phenomenology of Organ Transplantation*. The paper emanates a paradoxical flavor of the presence of alterity, the alien or otherness, in the self.

> Thus the foreignness of the grafted liver is less and less focused. The body itself has become a constant, ongoing source of foreignness altering itself as in echo, touching every sphere of my waking life... Transplantation is never in the past, ..it produces an inflexion in life that keeps an open reminder from the trace of the scar altering my settledness, bringing up death's trace. It is my horizon, an existential space where I adapt slowly, this time as the guest of that which I did not arrange, like a guest of nobody's creation (p. 270-271).

Along with the presence of the alien in the self and the inside in the outside, e.g., Varela's body captured on medical film as well as life and death decisions mediated externally by a team of medical experts, there is another level of paradox buried deep in Varela's paper. This level is implicit in any earnest effort by consciousness to seek its own truth, whether at the beginning, middle or end of life. It is the sobering recognition that all attempts to seek internal truths are fraught with lies.

In a 1975 paper, *Introductory Comments to Francisco Varela's Calculus For Self-Reference*, Howe and von Foerster note that Kant placed the autonomy of the observer at the center of his philosophy. This represented not so much a shift to subjectivity but instead to the ethics and responsibility that come with autonomy and self-creation. They state, "Lying and not objectivity is the problem and the force of the paradox of the Cretan liar. With his calculus of the paradoxical, the self-referential, the autonomous, Varela has opened for the first time the possibility of a Calculus of Responsibility" (p. 3).

In applying this calculus of responsibility to our selves, we can see how lies function in service of truth telling, at least in the realm of self-reflection. Whitehead (2001) notes that the ability to attribute false beliefs to others underlies tactical deception, purposeful lying, games like hide-and-seek, and the capacity to understand children's tales such as why Little Red Riding Hood gets into bed with the wolf. He also observes that the capacity to solve false belief tasks, which usually develops around four years of age, is essential to normal social interaction and may never develop in autistic children. In the same vein he writes that for young children pretend play is critical for learning to infer the mental states of others. Such play, which usually begins by around 12 months, can be understood as harmless lies harnessed in service of the capacity to tell the truth about the inner life of others, and ultimately of oneself as well. Pretend play is also the obvious precursor to making art. No wonder Shakespeare said, "The truest poetry is the most feigning," while Picasso's declared art "a lie that tells the truth." And Umberto Eco argues for deception as the precursor to signification.

When it comes to everyday life we tend to think of ourselves as stable, bounded entities, both psychologically and physically. We believe, or at least operate as if we believe, that we can make clear distinctions between what is inside and outside, me and not me, self and other. Yet these are truths based on a lie, a failure to recognize the structural underpinnings of identity as open as a fluid, dynamical, ever-moving, ever-renewing, ever-elusive processes with shifting boundaries that are infinitely deep and impossible to resolve.

Sadly, Varela's final paper went to press just as he died. Here are his final words, written under the subtitle, *Inconclusion*:

> I can see it: all of us in a near future being described as the early stages of a mankind where alterity and intimacy have been expanded to the point of recursive interpenetration. Where the body techné will and can redesign the boundaries ever more rapidly, for a human being which will be 'intrus dans le monde aussi bien que dans soi-meme' [extruded into the world as far as into himself] as the epigraph says. Even if my own window is narrow in time and fragmented in understanding. Somewhere we need to give death back its rights (p. 271).

Although Varela expressed this anomalous awareness in his twilight hours we all, whether in sickness or in health, must travel great distances away from ourselves in order to achieve intimacy within. Only as we gain the metaphorical capacity for objectivity, or distance from our selves, do we gain the capacity for intimacy. Yet paradoxically, no amount of distance allows us to perfectly achieve this goal. The paradoxical idea that intimacy with ourselves is a function of psychological distance, expressed with such anguish by Varela at the end of life, is beautifully captured in a more positive light by Theodore Roszak, who wrote, "the mind is gifted with the power of irrepressible self-transcendence. It is the greatest of all escape artists, constantly eluding its own efforts at self comprehension." Here we see the great paradoxical nature of psychological reentry at work. The very attempt to re-connect with our core is intrinsic to the very failure of this enterprise. And this failure itself represents its own ultimate success, that of self-transcendence.

References

Aurobindo. (1972). *On himself.* Pondicherry, India: All India Press.

Baldwin, J. (1902). *Social and ethical interpretations in mental development.* New York: Macmillan. (Original work published in 1897).

Barnes, H.B. (2000). *Stages of thought: The co-evolution of religious thought and science.* New York: Oxford University Press.

Beck, D. & Cowan, C. (1996). *Spiral dynamics: Mastering values, leadership, and change (developmental management).* Cambridge, MA: Blackwell.

Bullock, A., Stallybrass, O., and S. Trombley. (1999). *The New Fontana dictionary of modern thought, 3rd edition.* New York: HarperCollins.

Cooley, C. (1968). *Human nature and the social order. The self in social interaction. Vol. 1: Classic and contemporary perspectives.* New York: Wiley & Sons. (Original work published in 1902)

Combs, A. (2001). *The radiance of being: Complexity, chaos and the evolution of consciousness; 2nd ed.* St. Paul: Paragon House.

Copleston, F. (1962). *A history of philosophy, vol. 1: Greece & Rome, part 1.* Garden City, NY: Image Books.

Fisher, K, & Harter, S. (1999). *The construction of the self.* New York: Guilford.

Flavell, J. H. (1963). *The developmental psychology of Jean Piaget.* New York: Van Nostrand.

Freeman, W. (1995). *Societies of brains: A study in the neuroscience of love and hate.* Hillsdale, NJ: Erlbaum.

Gebser, J. (1949/1986). *The Ever-Present Origin.* (N. Barstad and A. Mickunas, Trans.). Athens, Ohio: Ohio University Press.

Green, E., & Green, A. (1977). *Beyond biofeedback.* New York: Delacorte Press.

Cook-Greuter, S., Miller, M., (1999). *Creativity, spirituality, and transcendence: Paths to spirituality and transcendence in the mature self.* New York: Ablex.

Hall, E. (1977). *Beyond culture.* Garden City, New York: Anchor Books.

Hampl, P. Review of *The best American short stories of the century,* L.A. Times, Book Review Section, July 18, 1999.

Happé, F. (1998). Understanding the self and others: Insights from autism. *The science of consciousness: Consciousness of the self, The 9th Mind and Brain Symposium Institute of Psychiatry,* London, November 14.

Howe, R. H. and H. von Foerster. (1975). Introductory comments to Francisco Varela's Calculus for self-reference. *International journal of general systems, Vol. 2,* pp. 1-3.

Kegan, R. (1982). *The evolving self.* Cambridge, MA: Harvard University Press.

Kegan, R. (1994). *In over our heads: The mental demands of modern life.* Cambridge, MA: Harvard University Press.

Kierkegaad, S. (1949). *The sickness unto death.* Trans. Walter Lowrie. Minneapolis: Augsburg Publishing.

Kohlberg, L. (1981). *Essays on moral development (Vol.1).* San Francisco: Harper & Row.

Laing, R. D. (1970). *Knots.* New York: Vintage Books.

Loevinger, J. (1976). *Ego development.* San Francisco: Jossey-Bass Publishers.

Mead, G. H. (1934). *Mind, self and society.* (C. W. Morris, Ed.). Chicago: University of Chicago Press.

Marks-Tarlow, T. (1999). The self as a dynamical system. *Nonlinear dynamics, psychology, and life sciences, Vol.3, No. 4,* pp. 311-345.

Nishitani, N. and R. Hari. (2000). "Temporal Dynamics of Cortical Representation for Action," *Proceedings of the national academy of science, USA, Vol. 97, No. 2,* pp. 913-918.

Palombo, S.R. (1999). *The emergent ego: Complexity and coevolution in the psychoanalytic process.* Madison, CT: International Universities Press.

Peirce, C.S. (1998). *The essential Peirce: Selected philosophical writings.* Bloomington, IL: Indiana University Press.

Prigogine, I. and I. Stengers (1984). *Order out of chaos: Man's new dialogue with nature.* Toronto: Bantam Books.

Putnam, H. (1990). *Realism with a human face.* Cambridge, MA: MIT Press.

Rizzolatti, G., Fadiga, L., Gallesi, V. and L. Fogassi. (1996), "Premotor cortex and the recognition of motor actions," *Brain research and cognitive brain research, Vol. 3, No. 2*, pp. 131-141.

Robertson, R. (1999). Some-thing from no-thing: G. Spencer-Brown's laws of form. *Cybernetics & human knowing, vol. 6. no. 4*, pp. 43-55.

Schuon, F. (1981). *Esoterism as principle and as way.* (W. Stoddart, Trans.). Middlesex, England: Perennial Books.

Singer, C. (1959). *A short history of scientific ideas to 1900.* New York & London: Oxford University Press.

Spencer-Brown, G. (1969). *Laws of form.* London: Allen and Unwin.

Spencer-Brown, G. (1979). *Laws of form* (rev. ed.). New York: E. P. Dutton.

Thompson, E. (2001). Empathy and consciousness. *Journal of consciousness studies, Vol.8, No. 5-7*; pp. 1–32.

Varela, F. (1975). A calculus for self-reference. *International journal of general systems, vol. 2*, pp. 5-24.

Varela, F. (1979). *Principles of biological autonomy.* New York: North Holland.

Varela, F. (2001). Intimate distances: Fragments for a phenomenology of organ transplantation, *Journal of Consciousness Studies, Vol. 8, No. 5-7*, pp. 259-271. Reprinted in E. Thompson, ed., *Between Ourselves* (Thorverton: Imprint Academic, 2001) [see back cover].

Varela, F., Thompson, E. and E. Rosch (1991). *The embodied mind.* Cambridge, M.A.: The MIT Press.

Varela, F., Maturana, H. R. and R. Uribe (1974). Autopoiesis: The organization of living systems, its characterization and a model. *Biosystems, Vol. 5, No. 4.*

Vygotsky, L. (1962). *Thought and language.* Cambridge, MA: MIT Press.

Wade, J. (1996). *Changes of mind.* New York: State University of New York Press.

Wilber, K. (1999). *Integral psychology.* Boston: Shambhala. Also published in the *Collected Works,* Vol.4; pp. 423-649. Boston: Shambhala.

Whitehead, C. Social mirrors and shared experiential worlds. *Journal of Consciousness Studies*, Vol. 8, No. 4, pp. 3-36.

Cybernetics & Human Knowing, Vol.9, No.2, 2002, pp. 49–63

Laws of Form and Form Dynamics

Louis H. Kauffman[1]

Abstract: This paper reviews the joint work of Francisco Varela and the author. The review is written as a chronology with emphasis on ideas and contexts.

I. Introduction

In the essay, I have tried to convey a picture of the form of my work and relationship with Francisco Varela. We wrote one paper together [3], a paper that he included as a chapter in his book "Principles of Biological Autonomy", and we had a long history of correspondence and mutual work and conversation over a period of years.

The first part of this essay (Section II) entitled "Remembering", is an historical account of our relationship and how it was linked with a mutual intellectual journey in relation to Spencer-Brown's Laws of Form. For this reason, I have included description and mention of a number of seminars and people that were crucial along this way. It was Francisco's understanding (and I assent readily to it) that mind is a conversational domain. The ideas that we articulated were part of a conversation that was ongoing with many others, and I have endeavored to mention a few of them in this essay. It is in the mind space generated by all us that these beautiful thought forms were born. The subject matter of the work was second order cybernetics and it occurred in a way that illuminates how a self-observing system can have self-awareness as a quality of the whole system. The notions of eigenforms and their dynamics in time is intimately related to this issue. In the remembrance I speak briefly of these structures and how they happened in the flow of our personal time. This section ends with a quote from Varela's last paper. In that paper he discusses the extraordinary event in which he receives a transplanted organ. In his philosophical reflection, that organ and the transplantation of thoughts and ideas from one mind/body to another are held at the same level. His courage of expression in this instance is beyond remarkable, and it focuses the key themes of his whole work.

In Section III I review the basics of Laws of Form, and in Section IV I discuss the concepts behind eigenforms and form dynamics. There is much more to say, but I hope that these last two sections will serve to elucidate the history of Section II. I really wish, on thinking through these themes once more, that I could send this paper over to Francisco and have a conversation with him about mutuality. I miss him and I miss his biological understanding of the cosmos.

[1] Department of Mathematics, University of Illinois at Chicago, 851 South Morgan Street, Chicago, IL 60607-7045. <kauffman@uic.edu>

II. Remembering

This essay is a remembrance, a remembrance of a friendship, a collaboration and a remarkable period of time. I met Francisco through our mutual interest in Laws of Form and the idea of an imaginary value. But lets start this story at the beginning.

I discovered the book "Laws of Form" [9] before I met Francisco. I encountered Laws of Form in 1974, some time after the book had been published and two years after I had completed a PhD in mathematics from Princeton University. I was teaching at the University of Illinois at Chicago Circle (as it was called at that time). I encountered a special experience in the foundations of thought and mathematics, almost as soon as I picked up that book by G. Spencer-Brown. Spencer-Brown's book was a turning point in my intellectual life. Laws of Form is a lucid exposition of the foundations of mathematics. It embodies a movement from creativity, to creation, to symbol, to system and language and thought and self. Expressing that creation took away the apparent ground of my previous conception of being, thought and understanding. There was no longer any distinction between the certainty or uncertainty of mathematics and the certainty or uncertainty of present experience. There was no longer any distinction between geometry/topology and logic. There was no longer any possibility that logic could be the foundation of mathematics, or that mathematics could have any foundation other than itself in the realm of experiencing of itself. There are many roads to this place. For me, Laws of Form came along at the right time. Later, I enjoyed reading accounts of similar experiences with Laws of Form by non-mathematicians such as Alan Watts and John Lilly.

A year after my encounter with Laws of Form a seminar arose at the Circle Campus (now called the University of Illinois at Chicago). This seminar, devoted to Laws of Form, met every Wednesday at the apartment of Kelvin Rodolfo, a professor of geology at the Circle. The seminar met cordially for the whole evening in a comfortable living room with food and drink and ample time for discussion. We were an eclectic group: David Solzman from Geography, Rachel MacKenzie, Gerry Swatez from Sociology, Joy Swatez, Kathy Crittenden from Sociology, Mike Lieber from Anthropology, Paul Uscinski a young computer scientist, Brayton Gray from Mathematics, myself and others. We read the book, argued over it, free associated to it, performed it and generally wandered in the space opened from the possibility of a single archetypal distinction. This seminar had a life of more than three years and it deeply influenced the lives of all its members.

Getting on into the book, Paul and I became fascinated by the recursive and circuitous world of Chapter 11. We invented for ourselves an interpretation of the workings of those circuits, and we found ourselves writing the reentering mark to express these ideas. It was with some glee that I discovered around this time an article in the Whole Earth Magazine about a young biologist, Francisco Varela, who had just written a paper [10] about Laws of Form and had worked out an algebra that included the reentering mark! We found Francisco's paper and added it to the discussion in the seminar. I resolved to get in touch with him.

A little research turned up Francisco's connection with Heinz von Foerster and the Biological Computer Laboratory at the University of Illinois at Urbana-Champaign, about 150 miles from Chicago. We had earlier in the seminar called up Heinz to tell him that we were studying Laws of Form. Heinz had written the brilliant review of Laws of Form that appears in the Whole Earth Catalog, where he characterizes it as "Spencer-Brown's transistorized 20th century version of Occam's razor". When we called Heinz and told him of our endeavor he laughed and we laughed in a joyous frenzy over the telephone.

I was fascinated by the notion of imaginary boolean values and the idea that the reentering mark and its relatives, the complex numbers could be regarded as such values. *The idea is that there are "logical values" beyond true and false, and that these values can be used to prove theorems in domains that ordinary logic cannot reach. Eventually I came to the understanding that this is the creative function of all mathematical thought.* But at that time I was fascinated by the reentering mark, and I wanted to think about it, in and out of the temporal domain.

The reentering mark has a value that is either marked or unmarked at any given time. But as soon as it is marked, the markedness acts upon itself and becomes unmarked. "It" disappears itself! However, as soon as the value is unmarked, then the unmarkedness "acts" to produce a mark!

You might well ask how unmarkedness can "act" to produce markedness. How can we get something from nothing? The answer in Laws of Form is subtle. It is an answer that destroys itself. The answer is that any given "thing" is identical with what it is not. Thus markedness is identical to unmarkedness. Light is identical to darkness. Everything is equivalent to nothing. Comprehension is identical to incomprehension. Any duality is identical to its confusion into union. There is no way to understand this "law of identity" in a rational frame of mind. An irrational frame of mind is (in this view) identical to a rational frame of mind. All is the working of the reentering mark. In Tibetan Buddhist logic there is existence, nonexistence and that which neither exists nor does not exist (See [16]). Here is the realm of imaginary value.

The condition of reentry, carried into time, reveals an alternating series of states that are marked or unmarked. This primordial waveform can be seen as

Marked, Unmarked, Marked, Unmarked,....

or as

Unmarked, Marked, Unmarked , Marked,...

I decided to examine these two total temporal states as representatives of the reentering mark, and I called them I and J respectively [2]. These two imaginary values fill out a world of *possibility* perpendicular to the world of true and false.

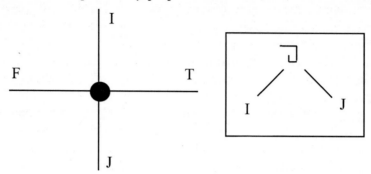

$$I = [T,F] <-----> TFTFTFTFTFTFTFTFTF...$$
$$J = [F,T] <-----> FTFTFTFTFTFTFTFTFT...$$

I wrote a paper [2] about I and J, showing how they could be used to prove a completeness theorem for a four valued logic based on True, False, I and J. I called this the "waveform arithmetic" associated with Laws of Form. In this theory the imaginary values I and J participate in the proof that their own algebra is incomplete. This is a use of the imaginary value in a process of reasoning that would be much more difficult (if not impossible) without it. Prior to that I had written a paper using Francisco's "Calculus for Self-Reference" to analyze the temporal behavior of self-referential circuits [1]. My papers were certainly inspired by Varela's use of the reentering mark in his analysis of the completeness of the calculus for self-reference that he associated with that symbol.

I also started corresponding with Francisco., telling him all sorts of ideas and recreations related to self-reference. We agreed to meet, and I visited him in Boulder, Colorado in 1977. There we made a plan for a paper using the waveform arithmetic. This became the paper "Form Dynamics", eventually published in the Journal for Social and Biological Structures [3]. An earlier attempt to publish it in the International Journal of General Systems was met by the criticism that we had failed to acknowledge the entire(!) Spanish School of Polish Logic. I still have the letter from that referee. Later I learned to appreciate Spanish Polish logic (a group of logicians in Spain working on DeMorgan algebras and related matters in multiple-valued logic). Francisco based a chapter of his book "Principles of Biological Autonomy" on form dynamics. I remember being surprised to find some of my words and phrases in the pages of his book!

The point about Form Dynamics was to extend the notion of autonomy inherent in a timeless representation of the reentering mark to a larger context that includes temporality and the way that time can be implicit in a spatial or symbolic form. *Thus the reentering mark itself is beyond duality, but implicate within it are all sorts and forms of duality from the duality of space and time to the duality of temporal forms shifted in time from one another, to the duality of form and nothingness "itself".* I believe that both Francisco and I felt that in developing Form Dynamics we had reached a balance in relation to these dualities that was quite fruitful, creative and meditative. It was a wonderful aesthetic excursion into basic science.

This work relates at an abstract level with the notions of autonomy and autopoesis inherent in the earlier work of Maturana, Uribe and Varela [11]. There they gave a generalized definition of life (autopoesis) and showed how a self-distinguishing system could arise from a substrate of "chemical" interaction rules. I am sure that the relationship between the concept of the reentering mark and the details of this earlier model was instrumental in getting Francisco to think deeply about Laws of Form and to focus on the Calculus for Self-Reference. Later developments in fractal explorations and artificial life and autopoesis enrich the context of Form Dynamics.

At the time (around 1980) that Francisco and I discussed Form Dynamics we were concerned with providing a flexible framework within which one could have the "eigenforms" of Heinz von Foerster [17] and also the dynamical evolution of these forms as demanded by biology and by mathematics. It was clear to me that Francisco had a deep intuition about the role of these eigenforms in the organizational structure of the organism. This is an intuition that comes forth in his books [12,13] and in his other work as well.

There is a more general theme that has been around since that time. It is the theme of "unfolding from a singularity" as in catastrophe theory. In the metaphor of this theme the role of the fixed point is like the role of the singularity. The fixed point is an organizing center, but it is imaginary in relation to the actual behaviour of the organism, just as the "I" of an individual is imaginary in relation to the social/biological context. The Buddhists say that the "I" is a "fill-in". The linguists point out that " I am the one who says "I"." The process that is living never goes to the fixed point, is never fully stable. The process of approximation that is the experiential and experienced I is a process lived in, and existing in the social/biological context. Mind becomes conversational domain and "mind" becomes the imaginary value generated in that domain. Heinz von Foerster [17] said "I am link between myself and observing myself."

The biological context is a domain where structural coupling and coordination give rise to mind and language. The fixed point is fundamental to what the organism is not. In the imaginary sense, the organism becomes what it is not.

Francisco invited me to participate in summer science seminars held at the Naropa Institute in Boulder Colorado in the early 1980's. We had a group of scientists and courses of lectures: linguistics (Alton Becker, Kyoko Inoue), poetry (Haj Ross), poetry and linguistics (Haj Ross), geography (David Solzman), biology(Varela and Maturana), psychology(Eleanor Rosch), Laws of Form (Kauffman), constructive mathematics (Newcomb Greenleaf) and more. We talked and talked. I do not know how many of us also meditated, but the atmosphere of the Buddhist Institute provided a wonderful place for the gestation and exchange of ideas.

After those Naropa years we saw each other a few more times . Once we drove together from a cybernetics meeting (a Gordon conference) to a weekend retreat at the Buddhist center Karme Choling in Northern Vermont. I saw him again in Paris in 1989 and once at at conference in Brussels ("Einstein Meets Magritte") in the 1990's.

In that time I kept returning to Laws of Form and our shared ideas [4,5,6,7,8].

It is clear on reading Francisco's last article [13] that he too remained always standing in the consideration of the embodiment and imagination of the self and the interplay of boundaries that it demands.

> So there it is: some two years ago I received the liver of another human being. An organ came tumbling down a complex social network from a recently dead body to land into my insides in that fateful evening of June 1. My sick liver was cut from its circulatory roots, and the new one snugly fitted in, replacing the vital circulation by laborious suture of veins and arteries. ...

> From this narrow window I must (we must) reflect on and consider an unprecedented event, that no accumulated human reflection and wisdom has ventured into. I take tentative steps, consider everything as only a tentative understanding, a lost cartographer with no maps. Fragments, no systematic analysis. We are left to invent a new way of being human where bodily parts go into each other's bodies, redesigning the landscape of boundaries in the habit of what we are so definitively used to call distinct bodies. Opening up the landscape where we can borrow a piece from another, and soon enough, order it to size by genetically modified animals. One day it will be said: I have a pig's heart. Or from stem cells they will graft a new liver or kidney and preselect the cells that will colonize what was missing in us, in a sort of permanent completion that can be extrapolated beyond imagination, into the obscene. This is the challenge that is offered to us to reflect on through and through, to live up to the challenge, to give us the insight and the lucidity to enter fully into this historical shift.

> ......

> In ten years, these reflections will probably be obsolete, the entire reality of transplantation having changed the scenario from top to bottom; all the work I must do is for a little window of history before it snaps out of focus and we are to re-start anew. [13]

There can ultimately be no distinction between bodily parts and parts of mind, parts of ideas, parts of the world that momentarily become possesions, or parts of the world that momentarily possess us. We derive our identity through relationship. Francisco's concerns were always for relationship. He had a wonder at the universe of autopoetic beings, always birthing, unfolding, interacting, dying and being reborn.

III. Laws of Form

Laws of Form [9] is a lucid book with a topological notation based on one symbol, the mark:

This single symbol is figured to represent a distinction between its inside and its outside:

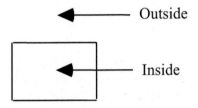

As is evident from the figure above, the mark is to be regarded as a shorthand for a rectangle drawn in the plane and dividing the plane into the regions inside and outside the rectangle. Spencer-Brown's mathematical system made just this beginning.

In this notation the idea of a distinction is instantiated in the distinction that the mark is seen to make in the plane. Patterns of non-intersecting marks (that is non-intersecting rectangles) are called expressions. For example,

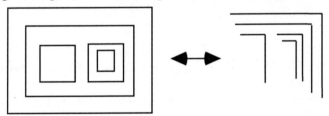

In this example, I have illustrated both the rectangle and the marks version of the expression. In an expression you can say definitively of any two marks whether one is or is not inside the other. The relationship between two marks is either that one is inside the other, or that neither is inside the other. These two conditions correspond to the two elementary expressions shown below.

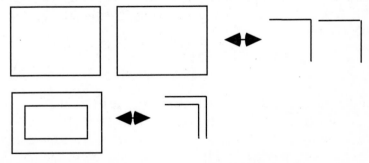

The mathematics in Laws of Form begins with two laws of transformation about these two basic expressions. Symbolically, these laws are:

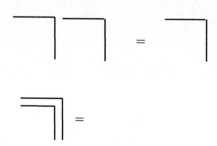

In the first of these equations, the law of calling, two adjacent marks (neither is inside the other) condense to a single mark, or a single mark expands to form two adjacent marks. In the second equation, the law of crossing, two marks, one inside the other, disappear to form the unmarked state indicated by nothing at all! Alternatively, the unmarked state can give birth to two nested marks. A calculus is born of these equations, and the mathematics can begin. But first some epistemology:

Spencer-Brown begins his book before this symbolism with a chapter on the concept of a distinction. "We take as given the idea of a distinction and the idea of an indication, and that it is not possible to make an indication without drawing a distinction. We take therefore the form of distinction for the form." From here he elucidates two laws:

  1. The value of a call made again is the value of the call.
  2. The value of a crossing made again is not the value of the crossing.

The two symbolic equations above correspond to these laws. The way in which they correspond is worth discussion. First look at the law of calling. It says that the value of a repeated name is the value of the name. In

one can view either mark as the name of the state indicated by the outside of the other mark! In this way the law of calling is instantiated in this equation.

In the other equation

we interpret the state indicated by the outside of a mark as the state obtained by crossing from the state indicated on the inside of the mark. Here the marked state is indicated on the inside and hence the outside must indicate the unmarked state. This gives rise to the equation, and it is a representation of the law of crossing. Note that the same interpretation yields the equation where the left-hand side is seen as an instruction to cross from the unmarked state, and the right hand side is seen as an indicator of the marked state. The mark has a double carry of meaning. It can be seen as an operator, transforming the state on its inside to a different state on its outside, and it can be seen as the name of the marked state. This combination of meanings is compatible in the interpretation that we have indicated here. The last equation embodies the compatibility in the perfection of identity that is not identity.

In this calculus of indications we see a precise elucidation of the way in which markedness and unmarkedness are used in language. For in language we say that if you cross from the marked state then you are unmarked. It is unambiguous in the realm of words. Not marked is unmarked. Here in this calculus of the mark these patterns are captured in a simple and non-trivial mathematics, the mathematics of the laws of form.

Paradox Lost

In Chapter 11 of Laws of Form, Spencer-Brown points out that a state that my appear contradictory in a spatial and timeless condition may reappear without paradox in a state of time. Just so with the famous paradoxes such as the Russell set of all sets that are not members of themselves. These are structures whose very definition propels them forward into the production of new entities that they must include within themselves. They are paradoxical in an eternal world and generative in a world of time. The simplest instance of such a description is the equation

$$J = \overline{J\;}\;\rceil$$

taken in the context of Laws of Form. For if J is equal to the mark, then the equation implies that J is equal to the unmarked state, and if J is equal to the unmarked state, then the equation implies that it is equal to the marked state.

$$J = \;\rceil \;\longrightarrow\; J = \overline{\rceil}\rceil \;=$$

$$J = \qquad\longrightarrow\; J = \;\rceil$$

There is no paradox when this form is seen to oscillate in time, but a new state has arisen in the form of the reentering mark J. In his first paper [10] on this subject, Francisco went back to the eternal point of view and decided to include a representative of the reentering mark at the basic arithmetical level along with markedness and unmarkedness. At this level the reentering mark would represent autonomy or autopoiesis. It represented the abstract concept of a system whose structure was maintained through the self-production of its own structure. This idea of a calculus for self-reference and the production of a symbol for the fundamental concept of feedback at the level of second order cybernetics captured many peoples imaginations, and it still does!

Here is the ancient mythological symbol of the worm ouroboros embedded in a mathematical, non-numerical calculus. The snake is now in the foundations and it is snakes all the way down.

One may argue that it is, in fact not appropriate to have the reentering mark at the very beginning. One may argue that it is a construct, not a fundamental entity. This argument will point out that the emergence of the mark itself requires self-reference, for there can be no mark without a distinction and there can be no distinction without indication (Spencer-Brown says there can be no indication without a distinction. This argument says it the other way around.). Indication is itself a distinction, and one sees that the act of distinction is necessarily circular. Even if you do not hold that indications must accompany distinctions, they do arise from them and the act of drawing a distinction involves a circulation as in drawing a circle or moving back and forth between the two states. Self-reference and reference are intimately interrelated.

In our work on form dynamics we place the reentering mark back in the position of a temporal construct, and in his book [12] Francisco even considers other methods of arriving at reentry such as the lambda calculus [15]. Even so, in biology one may view autonomous organisms as fundamental and one may look to see how they are generated through the physical substrate. It is a mystery that we face directly that the world that we know is the world of our organism. Biological cosmology is the primary cosmology and the world is fundamentally circular.

IV. Reentry, Eigenvalues and Form Dynamics

Consider the reentering mark (see section I of this paper).

This is an archetypal example of an eigenform in the sense of Heinz von Foerster [17]. What is an eigenform? Well an eigenform is a solution to an equation, a solution that occurs at the level of form, not at the level of number. You live in a world of eigenforms. What? You thought that those forms you see are actually "out there"? Out where? It has to be asked. The very space, the context that you regard as your external world is an eigenform. It is your organism's solution to the problem of distinguishing itself in a world of actions.

The shifting boundary of the Myself/MyWorld is the dynamics of the form that "you" are. The reentering mark is the solution to the equation

$$J = \overline{\quad J \quad}$$

where the right-angle bracket distinguishes a space in the plane. This is not a numerical equation. One does not even need to know any particularities about the behaviour of the mark to have this equation. It is more akin to solving

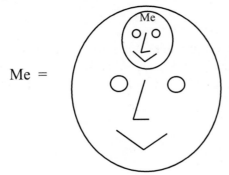

by attempting to create a space where "I" can be both myself and inside myself, as is true of our locus psychologicus. And this can be solved by an infinite regress of Me's inside of Me's.

Just so we may solve the equation for J by an infinite nest of boxes

Note that in this form of the solution, layered like an onion, the whole infinite form reenters its own indicational space. It is indeed a solution to the equation

The solution in the form

is meant to indicate how the form reenters its own indicational space. This reentry notation is due to G. Spencer-Brown. Although he did not write down the reentering mark itself in his book "Laws of Form", it is implicit in the discussion in Chapter 11 of that book.

Now you might wonder many things after seeing this idea. First of all, it is not obvious that we should take an infinite regress as a model for the way we are in the world. On the other hand, everyone has experienced being between two reflecting mirrors and the veritable infinite regress that arises at once in that situation. Physical processes can happen more rapidly than the speed of our discursive thought, and thereby provide ground for an excursion to infinity.

On the other hand, forms do not just build up. They also decay.

In Laws of Form (as we saw in Section 3 of this paper) we have the equation

where the nothing on the right hand side of the equals sign literally means nothing. Living in this context we see that the finite approximations to the reentering mark will oscillate between the values marked and unmarked, just as we described in Section 2:

This means that we now have two views of the reentering mark, one is purely spatial — an infinite nest of enclosures. One is purely temporal — an alternating pattern of marked and unmarked states.

All sorts of dynamics can occur in between these two extremes and this was the subject of the paper "Form Dynamics" that Francisco and I wrote together.

Note also how we can now see the relationship between the eigenform of the reentering mark and the two temporal forms of oscillation I and J (as discussed in Section 2). We can represent I and J as the two ways of observing the oscillation

$$...\text{MUMUMUMUMUMU}...$$

$$I = [M,U] \qquad [U,M] = J$$

where M and U stand for "Marked" and "Unmarked" respectively.

We define crossing (or negation) of a form shown as an ordered pair by the formula

$$\overline{[A,B]} = [\,\overline{A}\,,\,\overline{B}\,]$$

In this formalism [A,B] can be regarded as one view of the temporal form ...ABABABABABABA..., while [B,A] is the other (phase-shifted) view of that form. With crossing defined by the above formula, we shift the phase of the waveform and cross each individual term. The result is a formalism where I (shown below) and J are invariant under the operation of crossing.

$$\overline{U} = M$$
$$\overline{M} = U$$
$$\overline{I} = \overline{[M,U]} = [\,\overline{U}\,,\,\overline{M}\,]$$
$$= [M,U] = I$$

In this way we solve the eigenform problems for the reentering mark without using infinite regress. We accomplish this by staying close to the temporal interpretation of the form.

These patterns of form dynamics form the skeleton for the description and consideration of many structures in cybernetics and science. Elaboration of the solution to eigenform equations leads to the structure of fractals and to a philosophy that extends the notion of eigenvalues in physics. See [6,7] for a discusssion of this point of view. In his book "Principles of Biological Autonomy" Varela uses the form dynamics and associated waveforms as a way to speak about general issues of autonomy and self-reference. He also explores other structures of networks and self-reference, using ideas from the Church-Curry lambda calculus, lattice theory and graphs to begin a mathematical sketch for an abstract organism.

V. Summary

There is a kind of blinding clarity about these simple ideas near the beginning of Laws of Form. They point to a clear conception of world and organism arising from the idea of a distinction. Nevertheless, if you follow these ideas out into any given domain you will be confronted by, perhaps engulfed by, the detailed complexities of that domain. The non-numerical mathematics acts differently in than in traditional numerical models. It acts as an arena for the testing of general principles and as a metaphor that can be used in the face of complexity. The mathematics developed in Form Dynamics and in Principles of Biological Autonomy is not directly equipped to answer detailed questions. It is equipped with a flexibility and with the capacity to itself be complexified to meet the needs of more specific applications.

Here is an example that may shed light on this aspect. Consider the mutual forms A and B such that

$$A = \overline{B} \qquad B = \overline{A}$$

Each form (A and B) includes the other inside itself, creating a timeless mutuality indicated below the equations for the two forms. This is the form of mutuality. We are all familiar with manifold details of interdependence, and in certain instances (a handshake, a mutual look, a moment of regard) we come back to the form.

These calculi of self-reference provide the starting point for a reconsideration of the form of the world.

It is just such a reconsideration that Francisco followed in the years after his initial encounters with Laws of Form. He followed it in his biology and in his work in Buddhism and in his kindly and gentle communications with colleagues and friends. There is an eternity and a spirit at the center of each complex form. That eternity may be an idealization, a "fill-in", but it is nevertheless real. In the end it is that eternity, that eigenform unfolding the present moment that is all that we have. We know each other through our idealizations of the other. We know ourself through our idealization of ourself. We become what we were from the beginning, a Sign of Itself [18] .

References

[1] L. H. Kauffman. Network Synthesis and Varela's Calculus, International Journal of General Systems 4,(1978), 179-187.

[2] L. H. Kauffman. DeMorgan Algebras - Completeness and Recursion. Proceedings of the Eighth International Conference on Multiple Valued Logic(1978),. IEEE Computer Society Press, 82-86.

[3] L. H. Kauffman and F. Varela. Form dynamics. Journal of Social and Biological Structures (1980), 171-206.

[4] L. H. Kauffman. Sign and Space, In Religious Experience and Scientific Paradigms. Proceedings of the 1982 IASWR Conference, Stony Brook, New York: Institute of Advanced Study of World Religions, (1985), 118-164.

[5] L. H. Kauffman. Imaginary values in mathematical logic. Proceedings of the Seventeenth International Conference on Multiple Valued Logic, May 26-28 (1987), Boston MA, IEEE Computer Society Press, 282-289.

[6] L. H. Kauffman. Self-reference and recursive forms. Journal of Social and Biological Structures (1987), 53-72.

[7] L. H. Kauffman. Knot Logic. In Knots and Applications ed. by L. Kauffman, World Scientific Pub. (1994) pp. 1-110.

[8] L.H. Kauffman and J. M. Flagg. The Flagg Resolution. (in preparation).

[9] G. Spencer-Brown. Laws of Form. George Allen and Unwin Ltd. London (1969)

[10] F. J. Varela. A calculus for self-reference. (1975). Int. J. Gen. Systems.

[11] F. J. Varela, H. Maturana and R. Uribe, Autopoesis - the organization of living systems - its characterization and a model. Biosystems 5 (1974) pp. 187-196.

[12] F. J. Varela. Principles of Biological Autonomy. The North Holland Series in General Systems Research, G. Klir editor (1979). Elsevier North Holland Pub. (Chapter 12 - Closure and Dynamics of Forms).

[13] F.J. Varela (2001). Intimate distances: Fragments for a phenomenology of organ transplantation. *Journal of Consciousness Studies, 8* (5-7), 259-271. Reprinted in E. Thompson, ed., *Between Ourselves: Second-person issues in the study of consciousness* (Thorverton: Imprint Academic, 2001), [see back cover, *C&HK*, this issue].

[14] H. R. Maturana and F. J. Varela, "The Tree of Knowledge – The Biological Roots of Human Understanding", New Science Library (1987).

[15] H. P. Barendregt. "The Lambda Calculus Its Syntax and Semantics", North Holland (1981 and 1985).

[16] T. Stcherbatsky. "Buddhist Logic", (1958) Mouton de Gruyter.

[17] H. von Foerster. "Observing Systems", Objects: Tokens for Eigenbehaviours, pp. 274 – 285. Intersystems Publications (1981).

[18] C. S. Peirce, "Collected Papers – II, p. 2.230 – 2.231, edited by Charles Hartshorne and Paul Weiss, Harvard University Press, Cambridge (1933).

Dragonfly at Jewel Lake

small being, beautiful
jeweled eyes glinting, sparkling
blue iridescence, shining
marked with black and gold
here, so very near
shimmering, suspended
in a luminous globe
that brings you from a time
I cannot know …

or is this a blur of wings
as you hover
while I wonder:
what you are,
whether you are aware
of yourself and my delight in you,
whether you choose to visit me
with your mystery

am I just
a curious human
an odd intrusion in your realm
wading naked in this evening lake
sun gleamed and aglow

nothing much to you,
while for me you are a call
to see self and other
coexisting in intimacy,
to sense the richness unknowable
that inspires an evocation of mystery
to englobe this joyous moment

but I let it be, I do not choose.
I let you be, whatever you are,
leave you to your world,
whatever this is,
and swim quietly away
the smooth cool silk
caressing my soul
as water and body touch
in a psychic space.

Yet you follow me,
with your kin
rustling wings flitting
and sudden gleams
around the moving island
of my head.

Thus I live once more
the human ecstasy
of living aware of living
in a comfort of belonging
myself suspended here,
connected with your sparkling being
and with all other earth beings
knowing and not knowing
tasting explanation and mystery
entwined in this luminous moment.

Cybernetics & Human Knowing, Vol.9, No.2, 2002, pp. 67–76

Francisco Varela (1946–2001): A working memory

Ranulph Glanville[1]

This paper is a memorandum of working with Francisco Varela on our joint paper "Your inside is out and your outside is in." It is intended to show how we worked together—something of the process and the mood. The paper that was the outcome may be found in the literature (Glanville and Varela 1981), but working notes and outlines, correspondence, and a condensation written some time after the paper are published here for the first time, together with a certain amount of commentary and context. In the quoted material, I have altered nothing save occasionally tuning the language (though I have retained Francisco's American spellings): the point of this paper is not to correct, extend or otherwise modify the argument, which we developed between 1977 and 1981(which I continue to believe has validity). This account is a tribute, an example, and a little piece of history.

I will write of working with Francisco: a working memory.

Francisco Varela and I wrote one paper together. For me, this was an important paper. It said something I felt needed to be said (concerning Spencer Brown, distinction logics and re-entry) that reflected both Francisco's interests and mine. At the time we met (1977, at George Klir's NATO funded conference "Applied General Systems Research"), Francisco was the bright young star of (second order) cybernetics. In contrast, although three months older than him, I had only recently completed my doctorate and was at the very beginning of my public involvement with cybernetics. I wanted to work with him because I felt there were ideas to be explored together. But the benefit of association with the star he was didn't escape me, either. And I liked the idea of a band of "Young Turks" working together at the foundations of this new cybernetics.

In the end, I think this paper was probably more important for me than for him. Heinz von Foerster, a longtime Varela friend and mentor never even knew we had written a paper together, until I told him recently. But, although I may delude myself, I don't think this was because Francisco had no interest in what we wrote (as quotes later in this text show). Rather, I believe that his interests and his personal network were, at the time we completed the paper (summer of 1979), already changing. He was moving away from a direct involvement in cybernetics, was deeply involved in Buddhism, and found himself feeling more at home in a French culture than in the Anglo-Saxon. He was also, I believe, asserting his own autonomy.

[1] CybernEthics Research, 52 Lawrence Road, Southsea, Hants PO5 1NY, UK.
Email: ranulph@glanville.co.uk

What were these ideas?

A major preoccupation in the cybernetic world at that time was self-reference. The name "Cybernetics of Cybernetics" indicates this—just as did the alternative name in David Hilbert's Meta Mathematics project (the mathematics of mathematics) so effectively terminated by Kurt Gödel's (1931) Incompleteness Theory. In axiomatic systems, it seems that we cannot handle self-reference without entering areas that pose great difficulties. Yet biological systems which are made of cells that reproduce themselves, clearly do manage to contain a complete and consistent self-description and self-build system within themselves, through which they reproduce themselves over the lives of their biological "hosts."

Francisco had worked on the problem of self-reference, a major theoretical problem raised by autopoiesis—for autopoietic systems, first demonstrated through a computer model, are systems that produce their own means of production (Maturana, Varela and Uribe 1972). This is not the occasion for an extended discussion: I hope merely to locate the ideas we worked from and with. This becomes even more important when autopoiesis is generalised to "autonomy."

Using Spencer Brown's quixotic distinction logic, Francisco developed a calculus for self-reference, intended (I believe) to deal with this problem: to give a mathematical and theoretical basis for systems such as autopoietic ones that, if they were to be seen as fitting in with the great tradition of mathematics and logic, needed precisely that device that classical logic denied them.

My work also concerned the notion of self-reference. I was interested to construct a world in which the distinction between each observer was maintained. In this world I accepted that each observer (being distinct) would observe differently. I wanted to construct a theory that allowed that we observe differently while, nevertheless, believing we observe the same thing.

Taking the notion of observation as crucial, I suggested a universe populated by self-observing entities that I called Objects (i.e., objects of attention), which could be observed by other entities as a sort of combination of both observer and observed. I suggested that every observable could be thought of in this way: as a self-observing entity observed by other entities. (It was in the act of self-observation that observability, identity/autonomy and time—leading to a relational logic—lay.) Hierarchical arrangements together were possible not only because the structure I proposed created a logic from the structure, but because they were made, nevertheless, of Objects, all of which were observables on the same level. This world is, in itself, non-hierarchical: hierarchy comes from the arrangement by an observer of his observations, such that one Object seemed to be inside another, etc. From this I developed notions of representation and of the destruction of fundamentals by the very acts of observation through which they were observed. I have written a little about this in columns in this journal and I won't go any further into this here, nor will I introduce the (minor) modifications I have made since so the Theory still fits within the developments in my thinking that have occurred since this time. I will just point out that both Francisco and I were interested in distinction, self-reference, compositions into "larger" wholes, and the problems of stopping (of fundamentals and universes).

***

We worked at the paper in the following way.

At Klir's conference in Binghampton we discussed the possibility of a joint paper. I expounded what I thought were our areas of common interest, and we threw ideas at each other for some time. We spent an evening, Francisco writing the following summary of our paper at this stage:

1 The basic idea of this article is to make explicit the notion of the unwritten cross in G Spencer Brown's sense.[2] This amounts to an investigation of the limits of the process of distinctions, either upwards ("the universe") or downwards ("the elementaries").
2 We agree that at such limits, both extremes become indistinguishable, and thus, the same. This gives the totality a Möbius strip quality.
3 Anything which shows such structure will appear, to an observer, invisible, and, thus, to render it visible, it must ignore such circularity, create levels, and thus attribute properties to every distinction.
4 We conclude by showing that, in fact, this phenomenon, in the large, repeats itself at every level of distinction, and is patently visible as its boundary.

At the end of Klir's conference we had explored and affirmed this agreed area of mutual concern: re-entry in the form of distinction systems.

***

I later typed this as a memo, organised and cleaned it up, and expanded it into an extended (and somewhat poetic/ritualistic) outline. It became the document we pondered and then worked from at our next meeting. Here it is:

The Inside is out and the outside is in....[3]

An outline explaining how the Arithmetic of Closure, and the Theory of Objects interrelate, and how Objects infer Closure just as Closure infers Objects. Also, that the point of re-entry, where Objects meet relationships closing, is the point where the Universe is Fundamental, and v.v..
1 Consider a (large) number of Objects.
2 Relationships can be observed to hold between them. These are predicates of the Objects relating Objects together (some in predicate roles to others), as seen by another Object, and are demonstrated to be re-entrant in the Arithmetic of

---

[2] I have come to believe that the unmarked cross is perhaps the most unresolved matter in Spencer Brown's calculus. Many of the criticisms I have made may be related to this.

[3] We were working from the Theory of Objects, which forms the main contribution of my PhD dissertation (Glanville 1975), and the Calculus of Self-Reference and the Arithmetic of Closure—written with Joseph Goguen—which were, perhaps, Francisco's central pieces on Spencer Brown's distinction logic.

Closure:[4] an infinite number of relationships can be made from only very few
Objects, since each observation will be different.

3   Since this interconnected re-entry, or closure, can be observed, it too is an Object
    (i.e., the property of Closure of relationships subtended between Objects is itself
    an Object).

4   Steps 2 and 3 above are an argument in extenso.

---

5   The relationships between Objects re-enter.

6   Re-entry is an Object

7   An Object re-enters (since re-entry is an Object).

8   Steps 5, 6 and 7 above are an argument in semantics or lexicography, that is, they
    are an argument in the formulation of the Arithmetic of Closure.

---

9   Every property that is seen of an Object is
    a) another Object
    b) not the Object itself.

10 Properties subtended from Objects (relationships) re-enter (by the Arithmetic of
    Closure).

11 All Objects are fundamental.

12 Since properties re-enter and properties are Objects, Objects re-enter.

13 Steps 9, 10, 11 and 12 above are an argument in intenso.

---

14 All observables are Objects.

15 Re-entry is an Object.

16 Objects re-enter.

17 If Objects re-enter, and re-entry is an Object, the Object of re-entry is an Object
    (BUT the Object is not re-entry, and re-entry is not the Object).

18 Steps 14, 15, 16, and 17 above are an argument by tautology.

---

19 All forms of argument—in extenso, in semantics, (i.e., by substitution), in
    intenso, by tautology (hence paradox)—indicate that the mechanism of
    predication between Objects implies the Arithmetic of Closure, and that those
    things from which the Arithmetic of Closure can form relations and predications
    are Objects. Furthermore, the Object that is re-entry through which the system
    closes is the same no matter whether the argument is in intenso or in extenso (i.e.,
    the point of re-entry is the point at which Objects are fundamental (i.e., infinitely
    small, indivisible) and the Universe (i.e., infinitely large, all-embracing)).

20 The universal is fundamental, and the fundamental is universal. Herein the final
    paradox of closure in Objects, of Objects in closure.

Or, to put it another way, when I accept that I have reached a fundamental—that there
are no more distinctions—I have drawn the re-entrant distinction "that there are no
more distinctions to be drawn." And, when I accept that I have reached the Universal, I

---

[4]   Which might also be known as the Maths of Lexicography.

accept that I have drawn all distinctions, by drawing the re-entrant distinction "that there are no more distinctions to be drawn."

***

We met again in Amsterdam about a year later, after the Amsterdam 'World Organisation for General Systems and Cybernetics' conference, to work on these ideas: the summer of 1978. We sat outside a café (still there) on Leidseplein and threw more ideas in the air—ideas that came about through contemplation of the memo we had earlier composed. This time we were fortunate not to have to interrupt our "play" (above): Annetta Pedretti kindly took notes so the flow was not broken.

At the end of this session we not only had our agreed theme, but had taken ideas each of us had played with, and put them together, finding a form through which to write. The crux of what we had to say depended on the argument that any system created by drawing distinctions requires a constant re-distinguishing to distinguish the mark from the value (the boundary from what it bounds—i.e. contains or excludes) (Glanville 1979).

***

We wrote the final paper in a simple and straightforward manner. I took Annetta's notes and my memories, and, with the earlier notes and the summary turned them into the short text "Your inside is out and your outside is in," complete with many references. Francisco then tuned the text and edited the references so that just two remained. Apart from Spencer Brown, there was only a Beatles' song: the paper's title comes from the track "Everybody's got something to hide except for me and my monkey" on the 1968 "White Album.' He wrote, *"...I propose that we use no references other than G Spencer Brown and the Beatles. This, I believe, accentuates the elegance of what we are saying in the poetic dimension which is the most convincing for this kind of kinky logic."*

Funded by the British Academy and the British Council, I presented our paper at George Lasker's 1980 "Applied Systems and Cybernetics" conference in Acapulco and in 1981 it was published in the proceedings. Annetta created a Möbius strip "scarf" for me, with the paper's title crochéed in to it. I wore it for the conference presentation in a darkened and refrigerated room that denied Acapulco's tropical sun.

We exchanged comments about the paper. Francisco wrote, *"I have read it, digested it, and found it marvelous. It really has a strange fascination on me, and makes me a bit dizzy like a Julesz random-dot stereogram. But you managed to get the flavor rather nicely. Well done."* And, later on, *"I am delighted that we have persisted long enough to make something out of this. It's fun!"*

I replied to him, *"I must admit that, on reading through the paper to write the abstract I was very amused by it. It really is a very funny paper. And I'm glad we got it together in the end: especially the discussions about it. We must do another some time, if either of us can find the appropriate topic."*

We never found that appropriate topic to bring us together again to work: and I am sorry (as one so often is when it is no longer possible) that we did not make more effort to create chances to work together. Looking back on our paper, I see it now as much more of a joint undertaking—in the writing, the material and in the valuing—than I had imagined. We worked together, and although I initially suggested this collaboration, the outcome was, in the end, a true meeting of minds working towards a joint discovery, hopefully equally rewarding to each of us.

***

A few months later, I was inspired to write a very terse, personal summary of the (already short) paper. It is published here for the first time, as a sort of tribute. Of course, both the condensation and the notes (above) may make a fuller sense if you can access the original paper, sadly unavailable for publication here. Again, the condensation is presented here essentially as I wrote it. Note that I have modified the Spencer Brown distinction mark so that the two ends are differentiated (one has a base). This is so that we can see which end is which, in the figures. In the last figure, the Möbius of "contained" distinctions, I have redrawn the figure from the original in a tidy form. But I have also included a blow up of a scan of the original: I really like the quality of this image.

Condensed Version (1981) of "Your inside is out and your outside is in"[5]

a   Boundaries do not distinguish insides and outsides—their value lies in the marks that are the boundaries.

b   If existence depends on distinction drawing, then boundaries are self-distinguishing.

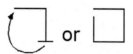

c   But, viewed by others, such boundaries appear to have intension and extension because the self-value of the self-mark cannot be attained by an other. Thus, the boundary appears to distinguish a value inside (or outside) it:

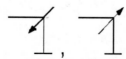

[5] Francisco was not party to the writing of this condensation, which I did on the spur of the moment one sunny morning during my regular train commute from London to Portsmouth.

d The distinction between mark and value can only be made by another observer by its continuous re-drawing in intension (or extension):

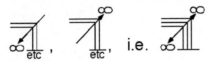

This can be seen as a source of (Piaget's) object conservation.

e The external differentiation of mark and value,

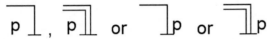

etc. is, in constant re-distinguishing, the source of time. Those distinctions which after a time appear indistinguishable one from the next (i.e.

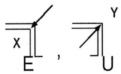

are what Heinz von Foerster calls the eigen behaviours of eigen objects. See von Foerster 1976)

f The apparent steadiness (in/out) in intension and extension resulting from the external observer leads to (unattainable) ideas of finality: the ultimate distinction of the elementary E (where, e.g., quarks become rishons),[6] the universal U. Since the value (to an other) is always inside (or outside) the ("final") mark, there is always another distinction: something "X" inside the smallest, "Y" outside the largest,

g This is equivalent to saying that the distinction we claim to be final always assumes, at least at the moment we declare it final, one more (post final) distinction—that it is final.

h Hence, distinctions being forms, at the unattainable final distinctions of intention and extension there is no formal difference, and thus there is re-entry.

i The analogue may be given that a self-distinction its like a Möbius strip. But seen from an other's point of view it seems to be a circle. Then, at the points of the final distinctions, the other discovers the re-entry of the Möbius strip, also. There is, at that point, no inside or outside. ALL is continuous; itself. There, everything shimmers.[7] Hence the way our paper (Glanville and Varela 1981) was written.

---

[6] I am not sure that this view is still held. But the point that all so far discovered indivisible fundamental parts have eventually been divided remains. I have argued this is not due to any properties of matter but rather to the way we examine and describe (Glanville 1980).

[7] A reference to a story that is contained in my cybernetics PhD (Glanville 1975).

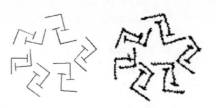

A Möbius of "contained" distinctions!

***

After we had finished this paper, I only saw Francisco once again: at the Gordon Research Conference on the "Fundamentals of Cybernetics" held in New Hampton, New Hampshire in the summer of 1984—although we did correspond occasionally. We sat on a raft in the middle of a lake, drinking beer and singing Captain Beefheart songs in the late, summer afternoon.

***

Soeren Brier has asked me what is the significance of this joint paper, and where it takes us? Although my intention has been to report on a collaboration, I will risk a little speculation.

1) Significance:

I think the significance of the paper was that it showed that, if you pursue the sort of logic Spencer Brown suggested (at least to us), you end up with a recursion that leads to re-entry (precisely because Spencer Brown's laws are laws of form), and that looking for absolute limits leads to circles. It was also, I believe, interesting to introduce the notion of a Möbius strip as the metaphor for the form of distinction, in place of the circle.

2) Where it takes us:

I think our paper brings out the central importance of notions of inside and outside in cybernetics. Using a spatial metaphor, do boundaries distinguish an in- and outside, or only themselves? This relates to one aspect at the crux of the debate about self-reference and second order cybernetics—what it allows us to say, bringing into contrasting view the circle and the Möbius strip forms as metaphors for different understandings, which gives rise to new understandings. In my case, I see this very spatially, in terms similar to those of Edwin Abbott in his (1884) "Flatland." You can either see the form of the distinction as a Möbius strip, which leaves us unable to speak of it (as an outsider), or as a circle, leaving us endlessly drawing distinctions—in intension and in extesion, never getting to that (presumed) self, but learning about endless observation. This is the view and the account of the other's view of that elusive (and presumed) self. So the paper is one of the initiators of the attempt to bridge the crucial and immensely difficult-to-handle difference between the knowing of the self, and the knowing of the other. Neither view is, I now believe, exclusively correct—we need both. As Francisco wrote about this time, "Not one, not two" (Varela 1976).

In other words, we distinguish the difference between in- and outside, self and other, and their views of these. Our paper, I believe, gives us hints of ways to recognise and work with these differences.

3) The bringer of comfort:

So, the importance of this work is that it helps us understand the world in which Spencer Brown's logic works. At the time Francisco and I worked on the paper, this world was still, for most, totally strange. Yet the declamations of the book such as "Draw a Distinction!" were compelling. It seemed necessary to explore the nature of the world of distinctions, and the form of this world. In our paper, I believe we pushed the envelope, and discovered this form. As it turned out, it bears a striking resemblance, in its re-entry, to the Möbius strip that I have claimed as the form for the distinction, and which Spencer Brown used on the cover of his later, limited edition (in 1994, and much to my delight). At the same time, others, especially Louis Kauffman in this journal, have explored the topology of distinction, of the conjoining of distinctions, and their meaning. All this could be said to rely not only on Spencer Brown's seminal work, but attempts such as in Francisco's and my joint paper to clarify the world that distinctions exist in. There is nothing much to develop from a paper such as this because its function is not as part of a developing argument, but to examine and set limits. As such, it is philosophical rather than mathematical: its function, if it has any, is to bring comfort and the feeling that we understand and know where we are, to provide legitimacy.

To tell us it's ok.

***

I did not report this episode primarily because of Francisco's brilliance, or to bring to the reader's attention a paper I believe is important. Rather, I would like to feel I have brought into view three points that help define and characterise Francisco for me.

Firstly, his generosity of spirit. In a very busy life, he made the time and maintained the enthusiasm to write this joint paper. This was, for him, common practice. A look at his publication list shows an exceptional amount of joint authoring.

Secondly, his delight in entering into the spirit of playful conversation, of batting ideas around till they eventually take on their own shape. This is a wonderful, synergetic way of working which we perhaps use too little.

Thirdly, his understanding of poetry, from which came his ability to deftly improve and tune what was written, and to exorcise what was inflationary and not entirely to the point (even in a foreign language).

I also wanted to communicate some of the pure delight we shared playing with our ideas together, watching them collide and grow. The joy and excitement of working at this level and in this way is something we rarely acknowledge—and that's a terrible pity.

Of course Francisco was brilliant, with an enormous skill as a theoretical and experimental scientist, his wonderful mind populated by an engaging curiosity and openness. But, I believe, he should be celebrated as a person for the qualities I mention above, even had he not made the important contributions he did as a scientist. These are the qualities that make us worthy humans, and they are qualities that go a long way towards creating elegant and precise work, helping us create beautiful understandings that make being human something really special. It is a pleasure to have come across Francisco's work. To have known him and worked with him is a greater pleasure.

Thank you Francisco.

References

Abbott, E (1884) Flatland: a Romance of Many Dimensions, London, Seeley & Co. republished in 1992 by Dover.

Beatles (1968) Everybody's got something to hide except for me and my monkey, in the White Album, Hayes, Apple-EMI.

Foerster, H von (1976) Objects: Tokens for (Eigen-) Behaviours, Cybernetics Forum vol. 8 nos. 3 and 4.

Glanville, R (1975) A Cybernetic Development of Theories of Epistemology and Observation, with reference to Space and Time, as seen in Architecture (Ph D Thesis, unpublished) Brunel University, 1975, also known as The Object of Objects, the Point of Points,—or Something about Things).

Glanville, R (1979) Beyond the Boundaries, in Ericson, R (ed.), Proceedings Society for General Systems Research Silver Jubilee Conference, London, London, Springer Verlag.

Glanville, R (1980) The Same is Different, in Zeleny, M (ed.) Autopoiesis, New York, Elsevier.

Glanville, R and Varela, F (1981) Your inside is out and your outside is in, in Lasker, G (ed.) Applied Systems and Cybernetics, vol. II, Oxford, Pergamon.

Goedel, K (1931) Ueber formal Unentscheidbare Saetze der Principia Mathematica und Verwandter Systeme, Monatshefte fuer Mathematik und Physik, vol. 38.

Maturana, H, Varela, F and Uribe, R (1972), Autopoiesis, University of Chile, Santiago.

Spencer Brown, G (1969) Laws of Form, London, George Allen and Unwin.

Spencer Brown, G (1994) Laws of Form (Limited Edition), Portland Oregon, Cognizer Co..

Varela, F (1975) The Calculus of Self-Reference, IJGS, vol. 1.

Varela, F. (1976) Not one, not two, CoEvolution Quarterly no. 12.

Varela, F and Goguen, J (1976) The Arithmetic of Closure, in Trappl, R et al (eds.) Procs. 3 EMCSR, Washington DC, Hemisphere.

Cybernetics & Human Knowing, Vol.9, No.2, 2002, pp. 77–82

ASC
American Society for Cybernetics
a society for the art and
science of human understanding

Varela's Contribution to the Creation of Cybersemiotics: The calculus of self-reference

Søren Brier[1]

The idea of evolution of living beings did not establish a firm foothold in the thinking of our culture until the 19th century. Evolution, though a biological concept, was nevertheless basically understood as a material change in body structure and function. In such a materialistic view great problems occur when one is trying to explain how mind came into being. How is it possible that the original "dead" world consisting of "pure" matter can foster living beings or observers with a sense of their own psychic existence?

In a modern biological view, the mechanics of life is deeply connected with the birth and development of the universe. Nevertheless, this materialistic cosmogony and evolutionary theory fails to explain the observer and the observing (the entire cognitive system.) After all, it is from within the observer and through language that the origin and progress of this evolution is explained. Our explanation takes place in the "praxis of living," as Maturana often puts it (e.g. Maturana, 1988).

In thermodynamics, cybernetics and especially in second-order cybernetics, the principle of self-organization is given the role of explaining evolution and the emergence of new qualities, such as life and mind. Even though self-organization is basic to the concept of autopoiesis, there is still a very long way from the dissipative structures of non-equilibrium thermodynamics to an understanding of living autopoietic systems. We have observed the spontaneous creation of organic molecules in experiments, we have Eigen's simulations of hypercycles with proteins and DNA, and we have spontaneous generation of cell-membrane-like structures and the autocatalytic chemical processes of Kauffmann (1995). But even with Hoffmeyer's further explicatory conditions for a system to be living

---

[1] This column has been initiated as a forum for commentaries by the Trustees of the American Society for Cybernetics. As a Trustee of the Society, Søren Brier <sbr@kvl.dk> contributes the following paper.

(Hoffmeyer 1985 and 1998), we are still unable to understand how the self-organization of matter can create living systems with a mind.

To a certain extent the notion of autopoiesis breaks with mechanical materialism through specifying the ability of living systems to self-organize their own molecules and cognition. This, in my view, takes life for granted. Next to nothing is said about ontological assumptions — about the medium in which organisms self-organize, and how life manifested during the development of the universe. Further, the problem of characterizing the force or forces that drive autopoiesis, is not addressed, proposing instead a "historical drift" along an epigenic course the dynamics of which are conserved through autopoisis and adaptation. More or less deliberately what I would consider a full philosophy is not attempted.

To a certain extent this is also the strategy of Heinz von Foerster. In some of his papers, he has nevertheless pointed out that there must be some sort of order and energy in the environment to establish an observing system and to have regular differences to observe (Brier 1996). This is a good point, but it does not penetrate into the deep connection between a living system and its environment and try to conceptualize the nature of that connection, except for his ideas of the ecological niche as a kind of "Umvelt" and objects as cognitive "eigen values" that seem to be purely psychological, or if not so, then he omits to declare the relevant metaphysics.

In his work on autopoiesis, Varela underlines that life and cognition are deeply connected. I take it as fundamental that the ability to make distinctions is basic to living systems, and I think that the first crucial distinction is between the system and its enviroment. One could say that this distinction creates the individuality of living systems. But how is this possible in a mechanistic worldview? Spencer-Brown, who is recognized as a profound mathematical philosopher contributing to second order cybernetics , instead clearly promotes an objective idealism in his book "Laws of Form"(1972). Here he speaks about how the world comes to see itself through making a distinction. Spencer-Brown, who Luhmann uses in his development of autopoiesis theory, thereby delivers an ontology based on self-organized closure that is compatible with autopoiesis and the whole idea of life and cognition, . Further, Peirce developed such an evolutionary objective idealism with the hylozoistic view, namely that matter contains rudimentary life "inside".

With the concept of autopoiesis, Maturana and Varela have taken an important step from the realm of biology in this direction , but more than a concept was needed. One of Francisco Varela's (1975) major accomplishments was with his calculus for self-reference which takes the work with autopiesis and Spencer-Brown's philosophy to a deeper philosophical level. Varela is clearly aware of the possibility of establishing a new and intimate connection between epistemology, logic and ontology. He sees the limitation of a dualistic view in explaning processes of self-organization, such as cognition. He introduces a third self-referential autonomous state. He writes :

The principal idea behind this work can be stated thus: we choose to view the form of indication and the world arising from it as containing the two obvious dual domains of indicated and void states, and a third, not so obvious but distinct domain, of a self-referential autonomous state which other laws govern and which cannot be reduced by the laws of the dual domains. If we do not incorporate this third domain explicitly in our field of view, we force ourselves to find ways to avoid it (as has been traditional) and to confront it, when it appears, in paradoxic forms. (Varela 1975, p. 19)

Varela aptly sees that self-reference goes beyond the mechanical laws. Peirce would talk about the "law of mind". "Life itself," I think Robert Rosen would have said. Varela further underlines the importance of the connection between self-reference and time, as he incorporates an important evolutionary view into this paradigm:

True as it is that a cell is both the producer and the produced which embodies the producer, this duality can be pictured only when we represent for ourselves a sequence of processes of a circular nature in time. Apparently our cognition cannot hold both ends of a closing circle simultaneously; it must travel through the circle ceaselessly. Therefore we find a peculiar equivalence of self-reference and time, insofar as self-reference cannot be conceived outside time, and time comes in whenever self-reference is allowed.[2] (Varela 1975, p. 20)

Through the formal introduction of time and self-reference, Varela introduces the connection between cognition and evolution and the arrow of time, without having to define it from the thermodynamic concept of entropy or its analogy to Shannon's concept of information, as Bateson did (Ruech and Bateson 1987). Thereby he establishes a view of cognition, which seems compatible with Pierce's semiotics.

Why is this important? Though leaving the objective view of information, autopoiesis theory and second order cybernetics have not yet developed a theory of meaning and signification, which connects the biological with the psychological cultural realm — Varela, Maturana, von Foerster, Glanville and Luhmann have taken steps in this direction. But C. S. Peirce, in his theory, has the same non-disciplinary, broad conceptual character as second order cybernetics and the same fundamental triadic and reflexive character that Varela has created for his autopoiesis theory, which we can now extend to second order cybernetics.

Pierce shows that a difference cannot become information before it has become so important to an observer/knower that he or she attaches a sign to it. Signs are what Varela uses to formulate and communicate his distinctions. Signs are needed for any formalized logic — as in Varela's calculus of self-reference. Indeed, only signs can be thought and communicated, and a difference, which cannot be communicated, hardly exists. A doctrine of signs and signification is needed.

In his semiotics, C.S. Peirce dived deep into the relation between mind, matter, natural laws, and the evolution of the universe. In accordance with modern thermodynamics and, to some degree, quantum field theory, Peirce sees the basic

---

[2]   I am here going back and further developing a specific point in my theory of Cybersemiotics, on which I first published in a conference proceeding (Brier 1993).

quality of reality as randomness or chaos. However, in order to explain how law and structure comes from randomness, Peirce (1892) finds it necessary to endow chaos with a particular quality, namely the tendency to form habits: to make distinctions that may last for a while. In this minimum statement he avoids saying too much about a virtual order in the transcendental, but he also avoids denying such an order. His purpose is to maintain an open boundary between physics and metaphysics.

In this way Peirce includes the observer and thereby life and mind. He sees the universe as arising from the random sporting of a concept that seems consistent with what the physicists now call the vacuum field, where suddenly a vibration or a wave passes the quantum threshold and becomes manifest. In accordance with modern thermodynamics this aberration expands, and in this process space-time is unfolded (as relativity theory sees it) and matter is created and organized into more complicated (self-organized) systems. However, the great difference is that he does not only see the Firstness vacuum as emptiness but also as fullness, as a hypercomplex dynamic process, which includes the characteristics of mind, matter and life. Peirce calls this pure spontaneity and pure feeling. He has a hylozoic viewpoint, namely he supports the doctrine that matter and life are inseperable. This means that matter posesses a type of life and sensation.

In his creation of a basic semiotic, Peirce did some very fundamental thinking regarding the necessary relationship between the subject, the sign or representamen and the object, and the minimum qualities he had to ascribe to them in order to create a realistic model of the process of knowing and signmaking. Peirce came to a triadic process-oriented view of the autonomous self-organization of signification and meaning. Peirce writes:

> A Sign, or Representamen, is a First which stands in such a genuine triadic relation to a Second, called its Object, as to be capable of determining a Third, called its Interpretant, to assume the same triadic relation to its Object in which it stands itself to the same Object. The triadic relation is genuine, that is its three members are bound together by it in a way that does not consist in any complexus of dyadic relations. That is the reason the Interpretant, or Third, cannot stand in a mere dyadic relation to the Object, but must stand in such a relation to it as the Representamen itself does. (Peirce 1955, p. 99)

I see Peirce's Interpretant as the observer of second-order cybernetics, which distinguishes differences. I think Peirce would agree that all living interpreters must be autopoietic systems — allthough he did most of his work with interpretants on higher levels. His concept of the interpretant goes deep into society, culture and history. In the continuation of the same text Peirce points out that the interpretant is itself a kind of sign, which also needs interpretation. What Peirce says is, that signification is never just a relation between a sign and its object. The sign can only signify what it is capable of being interpretated as. Therefore the interpretant is a necessary part of the sign. In accordance with Bateson we would say that we interpret differences as signs when the difference matters to us.

Peirce's reflexive or second-order definition of the interpretant is directed towards culture, history and the never-ending search for truth and knowledge. It underlines habits and historical drift as the social constructure of meaning, in agreement with Maturana and Varela. It is a drift, not a causally law-determined event, whose endpoint is mathematically calculable. In Perice's thinking it is pure feeling and life that evolves through the semiotic process.

Peirce's definition of signs is very cybernetic and self-organized. Thus it is *the semiotic web* that creates meaning. It is even so reflexive that it is second-order. Maturana and Varela's description of the autopoetic system, how it organizes itself in the historical drift of interaction in the domain of living, supports and supplements Peirce's description of the interpreter and its developmental relationship with culture — what Peirce calls *unlimited semiosis.*

A great part of our communication and thinking is not of our own doing; it is biological evolution and cultural history signifying through us. Therefore we will also have to accept that we have no ultimate control over thinking and communication — in short, signmaking. The human being must to a certain extent accept himself as the "place" where speech is created and a person is, in part, self-created through the social process of languaging.

The understanding of our basic situation as knowing beings without full conscious control over the creative aspects of language, and no conceptual scientific insight into any kind of timeless logic behind all language, is an entrance to understanding logic and rationality in another way. Signification, meaning, rationality and logic are not born fully-fledged but are gradually crystallized out of vague beginnings through the historical drift of praxis and the dance of languaging, the intimate development of the "life forms" and "language games" that Wittgenstein described. Furthermore, we must accept signs and concepts as just as fundamentally a part of reality as material objects. It is in accordance with evolutionary thoughts that we must realize that vagueness is first, precision is second and understanding is third. This is a view of semantics, symbols and logic that is differing from that of mainstream cognitive science and artificial intelligence (AI), the so-called "symbolism" that is often criticized by researchers within second order cybernetis and autopoieses theory.

Varela adopted a Buddhist view of the world, and therefore his view came much closer to Peirce and Spencer-Brown's objective idealistic view than that of Maturana.Thus, it is Varela's work in the calculus for self-reference that is the crucial bridging step in the development of this new field where autopoiesis theory, second order cybernetics and semiotics merge into Cybersemiotics. The work is further developed in his enaction theory. Varela was a fine man and an independent thinker as well as a creative scholar.

References

Maturana, H. (1988): "Ontology of Observing: The Biological foundation of Self Consciousness and the physical Domanin of Existende." In Donaldson, R.E. (ed) (1988): Conference workbook for "Texts in cybernetic Theory." An In-Depth Exploration of the Thought of Humberto Maturana, William T. Powers, Ernst von Glaserfeld. A conference of The American Society for Cybernetics, October 18-23, Felton, California, pp. 4-52.

Ruech, J, Bateson, G. (1987): *Communication: The social matrix of psychiatry,* New York. Norton Cop.

Brier, S. (1993): "Cyber-Semiotics: Second order cybernetics and the Semiotics of C.S. Peirce" in Proceedings from the Second European Congress on System Science, Prague, October 5-8. 1993. AFCET.

Brier, S. (1996): "From Second-order Cybernetics to Cybersemiotics: A Semiotic Re-entry into the Second-order Cybernetics of Heinz von Foerster", *Systems Research*, Vol. 13, No. 3, pp. 229-244, 1996

Hoffmeyer, J (1995): "The Swarming Cyberspace of the Body". *Cybernetics &Human Knowing* 3(1), 16-25.

Hoffmeyer, J (1998). "Surfaces inside Surfaces", *Cybernetics & Human Knowing* 5 (1), 33-42.

Kauffman, S. (1995): "*At Home in the Universe*." Oxford University Press.

Peirce, C.S. (1955): *Philosophical Writings of Peirce: Selected and Edited with an Introduction by Justus Buchler*, New York Dover Publications.

Peirce, C. S. (1892): "The Doctrine of Necessity Examined," *The Monist*, Vol. II, No. 3, April 1892.

Spencer-Brown, G. (1972): "*Laws of Form*, 2nd edition, New York.Varela, F. (1975): "A calculus for self-reference." *International Journal for General systems*, Vol. 2, pp. 5-24.

Cybernetics & Human Knowing, Vol.9, No.2, 2002, pp. 83–91

The Embodied Mind: Cognitive Science and Human Experience[1]

A book review and commentary

Allan Combs, Guy Burneko, Sally Goerner, Tracy Brown and Herbert Guenther

Allan Combs, Systems Theorist:

The following commentaries on this book were originally written at the time of its publication.[2] They are re-edited for inclusion in this commemorative issue as a kind of snap-shot of the many impressions that the book made on scholars of the time. As can be seen, these range from profoundly respectful to relatively indifferent, and even to negative, in the case of Professor Herbert Guenther's assessment of the Buddhist aspect of the book's philosophical orientation. We hope you will find these commentaries of interest.

As we see in the present pages, Francisco Varela was a many-sided man. He had the rare ability to move easily between refined theoretical reflection and the concrete and complex business of operating a neurobiological laboratory in Paris. He also moved between cultures with ease; and he is honored for his spiritual commitment. The latter led him not only to become an accomplished Buddhist practitioner, but to play a seminal role in the creation of a series of dialogues between leading Western scientists and the Dalai Lama (Bstan-Dzin-Rgya-Mtsho, Hayward, Varela, 2001; Dalai Lama, Engel, Gackenback, & Varela, 1997). Perhaps his most lasting contribution was made early in his career while working with Humberto Maturana, when the two of them developed the notion of *autopoiesis* (Maturana & Varela, 1980, 1987; Varela, Maturana, & Uribe, 1974), the idea that biological as well as cognitive systems can be understood as ongoing *self-creating* processes that define their own identity by conserving their structure while exchanging energy and information with the environment, with which they share a structural coupling.

[1] by Francisco J. Varela, Evan Thompson, Eleanor Rosch, Cambridge, MA: The MIT Press, US $25.00 hardcover, 308 pp., includes figures, appendices, notes, references, index, 1991.

[2] *World Futures: The Journal of General Evolution.* 37 (4), 219-225.

With the publication of *The Embodied Mind,* with Evan Thompson and Eleanor Rosch, however, Varela became known for the first time to a much wider audience of cognitive and brain scientists who found in this seminal volume a broad mapping of relationships between virtually every branch of the burgeoning cognitive sciences of the early 1990s and his own ground-breaking conceptualization of the brain and its action in relationship to consciousness and behavior. Essentially, Varela, Thompson, and Rosch rejected the computationalist agenda, then dominating the cognitive neurosciences, and systematically argued for the idea that the organism is structurally coupled with its environment in an intimate ongoing exchange of energy and information that gives birth to consciousness and cognition as emergent features of that interaction, and not as separate entities to be accounted for by traditional dualisms (mind-body or organism-environment dualisms). Thus, he put consciousness and behavior on a radically different ground, and prepared many readers to anticipate the importance of the neural network views of the brain that were just becoming wide spread at the time of this writing.

As an evolutionary biologist, this line of thinking also allowed him to develop the important notion of "natural drift," by which organisms, structurally coupled with their environments, undergo gradual evolutionary change that is not "driven" by hard demands of survival, but rather only constrained by them. In other words, to drift into new and creative forms that are not determined by the environment, but only limited by it to the necessities of living.

Guy Burneko, Philosopher:

Continental philosophy has been in the air for some time; and the insights of Buddhist theory and practice – at least since Christmas Humphreys, Aldous Huxley and Alan Watts – have been widely accessible for a good couple of generations; no less, with the work of Ilya Prigogine, Elizabeth Stengers, Erich Jantsch, James Gleick and others, has nonlinear dynamics become a part of our intellectual repertoire in recent years. Never, however, has their hybrid vigor been so amply demonstrated as in this sustained argument by Varela, Thompson, and Rosch, which is notable both for its strong frontier reach and the capable grasp conveyed by its remarkably clear and well-organized composition.

Herbert Guenther's meticulous *From Reductionism to Creativity: rDzogs-chen and the New Sciences of Mind* (1989) will prove to be more minutely informative on technicalities of Buddhist thought in connection with self-organization dynamics, as may Jacques Derrida, say, or Gregory Bateson on the "difference that makes a difference;" but no one has to my knowledge yet – even Jeremy Hayward – so lucidly and with such measured grace of prose and of reason conjoined the multifold axes of nonfoundational, experiential, and systems theoretical approaches to mind and evolution as have the authors of *The Embodied Mind.* Any fault it may have, and I'll come to that in a moment, is merely the excess of its virtue; it is an important essay; and it honors the members of the generation in, and from, which it appears. With this, we clearly enter the twenty-first century: the first of a genuinely planetary and transdisciplinary discourse; one distinctively informed by experiential and not solely cerebral intelligence.

In a nutshell, the argument progresses from the claims of Maurice Merleau-Ponty, shared by the authors, that embodiment "encompasses both the body as a lived experiential structure and the body as the context . . . of cognitive mechanisms;" mind is known both objectively and from the immediacy of our experience. The West, having no "disciplined manner" by which to practice a "hands-on, pragmatic approach to experience" will benefit from this text as a "bridge between mind in science and mind in experience . . . a dialogue between . . . Western cognitive science and Buddhist meditative psychology."

Moreover, in the enactive program presented in this book, the authors "explicitly call into question the assumption, prevalent throughout cognitive science, that cognition consists of the representation of a world that is independent of our perceptual and cognitive capacities by a cognitive system that exists independent of the world." They outline instead, "a view of cognition as *embodied action* in the context of evolution understood not as "optimal adaptation" but as what is called "natural drift." The conclusive point, that "cognition has no ultimate foundation or ground beyond its history of embodiment" is that whereon converge the Buddhist, Continental and self-organizational premises. Analogically speaking, this is the Copenhagen Interpretation of consciousness with an Eastern evolutionary accent.

Yet it rings true each careful step of its exposition. There are historical preparations (on cognitivism, connectionism, AI and emergentism), philosophical excurses and clarifications (on nonfoundationalism, Abhidharma, philosophical hermeneutics, object relations and Cartesianism), and there are technical descriptions for purposes of precise illustration (e.g., on brainwave periodicity and on color perception). The ethics of self-organizational "Laying Down A Path In Walking" are developed at the end, as are the finesse of Nagarjuna's "Middle Way" and the skill of mindfulness/awareness elsewhere.

In fact, though a shortish book, it is almost a curriculum in itself. The one thing that may mar it is what seems a needlessly continued treatment of nihilism in the West. However, we may gladly concede the outcome of this treatment – for it reveals that "nihilism is in fact deeply linked to objectivism": an idea whose development by these writers is itself worth the price of admission.

In closing this brief comment, I would like to single out the contribution in particular of the principal author, Francisco Varela. His work is well-reputed internationally and continues as a benchmark of contemporary scholarship for its interdisciplinary and intercultural address. Co-author with Humberto R. Maturana of the earlier *The Tree of Knowledge: The Biological Roots of Human Understanding* (Shambhala, 1987), Varela has been a Lindisfarne Fellow in the Program in Biology, Cognition, and Ethics, a student of Chogyam Trungpa, is

Director of Research at the *Centre National de Recherche Scientifique* and Professor of Cognitive Science and Epistemology at the Ecole Polytechnique in Paris; he has, as well, been an associate of the Naropa Institute in Boulder, Colorado (USA), where began the inspiration for this book.

He, with Thompson and Rosch, has now given us another ripe fruit of the tree of mind, one indication of whose maturity is that it has outgrown and superseded the "cloven fiction" that once hedge us epistemologically. In the words of the three:

> Insistence on the codetermination or mutual specification of organism and environment should not be confused with the more commonplace view that different perceiving organisms simply have different perspectives on the world. This view continues to treat the world as pregiven; it simply allows that this pregiven world can be viewed from a variety of vantage points. The point we are making, however, is fundamentally different. We are claiming that organism and environment are mutually enfolded in multiple ways, and so what constitutes the world of a given organism is enacted by that organism's history of structural coupling. Furthermore, such histories of coupling proceed not through optimal adaptation but rather through evolution as natural drift.

> The treatment of the world as pregiven and the organism as representing or adapting to it is a dualism. The extreme opposite of dualism is a monism. We are not proposing monism; enaction is specifically designed to be a middle way between dualism and monism.

Having already called attention to this book's artful composition and to its merit, even alluding to its significance by having made a comparison between it in consciousness studies and the advent of quantum theory in physics, I will finish by suggesting that we have not had so wonderfully contrived a perspective since that of John Scotus Eriugena in his own time so long ago.

Sally Goerner, Systems Theorist:

Before getting into the meat of this review, let me give my concise recommendation: you should read this book because you will go a far piece before finding a more dazzling synthesis of cognitive science, artificial intelligence, epistemology, continental philosophy, and Buddhist philosophy (plus a dash of psychoanalytic theory). The book is well-written, the descriptions are remarkably accessible, and the examples and supporting observations are plentiful. If nothing else, reading this book will make you think.

The book's topics are those most thorny of all subjects, how to understand "mind," "self," "cognition," "consciousness," and the human experience. The authors' goal is to establish a dialogue between modern cognitive science (including computational models of mind) and Buddhist meditative psychology, while at the same time showing their relationship to 20th century continental philosophy, particularly that of Heidegger, Merleau-Ponty, and Nietzsche. Toward this end they produce some beautiful summaries of the various Western scientific approaches to the study of mind/intelligence. The best part of this book is the overview of the development of Western cognitive science in the last 20 years – you will find everyone from Minsky and the connectionists, to Walter Freeman's neurophysiological insights, to J.J. Gibson and the ecological approach.

However, this book does have drawbacks. It is marred by a tendency to make caricatures of other attempts at understanding, and to automatically discount objections by casting the opposition as "in denial" or "resistant." What might have been very interesting arguments are often made weak by a tendency to shift between different uses of the same term, often within the same argument and often in a way that seems bent more on setting up a straw man than on increasing understanding. For example, belief in an external world is sometimes used in the naïve sense that there is something more than just our own mind, sometimes in the sense of a world that is completely independent of the observer, and sometimes as a world that is both completely independent and static or prefixed. The result is a tone that comes across like a negative campaign ad – one fears its impact will have more to do with its inflammatory nature that its quality.

All of this I find deeply unfortunate. The authors have taken on a tremendous and important task. And they have a significant positive contribution to make. When they talk about their own vision of enactive cognitive science they hit their stride. Here the argument is clear. Enactive cognitive science "emphasizes the growing conviction that cognition is not the representation of a pregiven world by a pregiven mind but is rather the enactment of a world and a mind on the basis of a history of a variety of actions that a being in the world performs" (p. 9). A couple of quotes will clarify:

> Cognitive science slowly drifted away from the idea of mind as an input-output device that processes information toward the idea of mind as an emergent and autonomous network. (p. 151)

> Knowledge depends on being in a world that is inseparable from our bodies, our language, and our social history – in short, from our embodiment. (p. 149)

Thus, cognition is not to be studied as inner versus outer worlds but rather as embodied action. Embodied action highlights the fact that:

> Cognition depends upon the kinds of experience that come from having a body with various sensorimotor capacities, and second, that these individual sensorimotor capacities are themselves embedded in a more encompassing biological, psychological, and cultural context. By using the term action we mean to emphasize once again that sensory and motor processes, perception and action, are fundamentally inseparable in lived cognition. Indeed, the two are not merely contingently linked in individuals; they have also evolved together. (p. 173)

The book is well worth the reading because of its breathtaking display of knowledge and its introduction of both evolutionary and ecological themes into cognitive science. Whether it has established a dialogue between cognitive science and Buddhist psychology remains to be seen.

Tracy Brown, Cognitive Psychologist:

As a cognitive psychologist trained squarely in the heart of the information processing paradigm, and with long-standing interests in linguistics, Eastern philosophical thought, and neuroscience, I approached *The Embodied Mind* with a curious mixture of anticipation, hopefulness, and skepticism. My anticipation was due to the heady mix of new and exciting topics and approaches discussed in the book, including the rapidly expanding area of connectionist theory in cognitive science, its relation to dynamical systems theory and chaos, an analysis of basic issues in cognitive science such as symbolic representation and realism in perception, and several other seemingly diverse topics such as color vision, evolutionary theory, and Rodney Brooks' work on subsumption architecture in robotics. My hopefulness was founded in a profound sympathy for the authors' basic objective, to explore the linkages among these approaches and issues in the light of a vigorously-developed Buddhist Philosophical stance, all in an effort to construct a dialogue between – or perhaps a synthesis or integration of – science and phenomenology. Despite the powerful and appealing nature of such a rapproachment between science and experience, I was skeptical that such an integration is possible, and that a new paradigm for science in general, and for cognitive science in particular, could be constructed from the effort. In part, this was because of my coming reluctantly to the conclusion, years ago, that the relationships between these modes of thought are inherently paradoxical, and that the only wise response to the paradox is to embrace it rather than to try in vain to resolve it. In approaching *The Embodied Mind* I was hopeful that I was wrong but afraid that I was right. After reading it, I am more hopeful about being wrong but far from convinced that a successful integration is achieved.

Perhaps the most basic theme underlying my response to *The Embodied Mind* is that many of the arguments involving the self, symbolic processing, representation, and perception – which the authors seem to feel are controversial and radical departures from established thought – are not nearly so radical as they believe. The concept of a self, or "I", which the authors convincingly dismantle, does not by my reading appear to be an indispensable component of information processing theory. In fact, explorations of automaticity, implicit learning, subliminal perception, and other forms of unconscious information processing have highlighted the idea of coordinated but autonomous systems, entirely congruent with their approach.

Their treatment of symbolic-level processing seems equally uncontroversial. To my thinking, the historical notion of symbols as units of processing was little more than a notational convenience during the inception of the cognitive approach, inspired by the digital computer and shaped by the need to talk about mental representation in the aftermath of behaviorism. In the increasingly intense and focused attempts to draw convergences between neurophysiology and cognition, the commitment to symbolic representation – in the sense that discrete symbolic units act as functional entities in the brain – has given way to more diverse theories about how information is represented both neurologically and computationally. The shift in focus seems apparent in the advent of semantic network theories of knowledge representation in the mid- to late- seventies, and, of course, in the emerging popularity of connectionist models in the past 10 or so years. In fact, it would appear that the source of most of the resistance to connectionist modelling within cognitive/information processing psychology comes not from a commitment to explicit symbolic representation, but rather from another class of theories known as episodic models which are similarly uncommitted to symbolic-level processing.

At a more general level, the same response applies to the authors' treatments of mental representation, perception, and evolution. In all of these areas, the authors carefully develop the theme of codependent arising by citing examples of enactive or interactive relations between the known and the knower, the perceived and the perceiver, the organism and its environment. Throughout these arguments, the authors seem to be pointing toward a new paradigm, a new way of approaching these topics – the "middle way," as they say – which solves the philosophical problems presented by the inherent dualism in these conceptualizations. While enthusiastic about the examples the authors use, and in sympathy with the enactive or interactive implications that the authors extract from their examples, I fail to see how their observations necessitate (or explicate) a new paradigm and how "the middle way" resolves the problem of dualism without also dropping entirely from the equation the terms known vs. knower, perceived vs. perceiver, and organism vs. environment. It seems to me that without these concepts the authors would not have been able to capture the emergent and dynamic properties which motivate their approach.

In summary, I found myself thinking "of course" to much of what is contained in this book. From my information processing perspective, the theoretical developments contained in it are not so radical as the authors appear to believe, and so while I am sympathetic with the concepts of enactive or embodied cognition, codependent arising, and dynamic/emergent systems that the authors develop, I am not inclined to view these ideas as the foundation of a new paradigm. Nevertheless, there is much of value in this book, particularly in its ambitious and challenging attempt to synthesize diverse topics in philosophy and science within the context of the Madhyamika tradition of Buddhist thought. I'm still thinking about that, and if the success of a book can be measured in terms of its tendency to stimulate thought, then this is a very successful book, indeed, and well worth a careful reading.

Herbert Guenther, Buddhist Scholar:

This book is a strange mixture of incongruities that perpetuate the myth of materiality and base themselves on an extreme reductionism. It begins with listing the almost countless opinions that make up what is called cognitive science – this magic word that pretends to say something, whilst actually saying nothing, simply because those who use this word do not realize that it is a contradiction in terms. It is one of the axioms of science that it must be objective. In order to be objective it must exclude anything that is subjective. Unfortunately, cognition is a very subjective phenomenon. Hence, in order to be objective the cognitive scientist ought to exclude cognition and, since this is not possible, in his endeavor to uphold the myth of materiality and to preserve the postulate of the separation between the observer and the observed in order to have something he can measure and experiment with, he reduces cognition to an epiphenomenon of organic processes. Worst of all, he does not realize that which he is looking for is precisely that which does the looking. Contrary to a widely held belief, science does not product "objective truth(s)," and its basic organizing principle, logic, is singularly incompetent in dealing with living experiences.

This decidedly unpromising start is not improved by dragging in what the authors believe to be Buddhism. There is no evidence that original sources, be they in Sanskrit or in their Tibetan translations, have ever been consulted. If they had, the authors would not have misspelled the title of what is considered the major work by Nagarjuna who is extolled beyond measure. I am afraid that this misspelling reflects their ignorance of the difference between *madhyamaka* and *mādhyamika* – the first naming a text, the second a follower of the line of thought developed in this text. Nor do they seem to be aware of the fact that Nagarjuna's writings reflect only one aspect of Buddhism, namely, an interest in logic, and that he was severely criticized for his reductionism by other logicians. It is significant that his way of thinking did not produce a single thinker of renown. Even some of the Tibetans were quick to note that his "middle" (Skt. *madhyama*, Tib. *dbu-ma*) was a static middle and had it superseded by a "dynamic" middle (*dbu-ma chen-po*) that has no beginning and no end, but always a middle from which it grows and which it overspills. This dynamic middle is direction in motion. For the rest, what the authors offer as Buddhism is verbose hearsay reporting with a couple of Sanskrit and Tibetan words which they do not understand, thrown in for good measure.

One wonders how such a dynamic concept as autopoiesis ever strayed into the jungle of outdated and static notions. Was it a Freudian slip which the authors quickly tried to cover up by not mentioning the name of authors like Richard M. Zaner, Maurice Merleau-Ponty and others who dealt with the problem of embodiment?

In the journal *Nature* its editor described Rupert Sheldrake's book *A New Science of Life* (1981) as "the best candidate for burning there has been for many years." Concerning *The Embodied Mind* I would not take this stand. I would follow Mammata, the author of the *Kāvyaprakāśa*, a work on poetics, who is reported to have said, when shown a copy of his nephew Śriharta's *Naiṣadhacarita*, that he regretted not having seen it earlier. It would have saved him the trouble to search for examples to illustrate the faults of poetry in many works, because they were all present in his nephew's work. In this vein I would say that *The Embodied Mind* is an excellent work because it brings together all the absurdities of Western "science" with its reductionist (logical) positivism and of Buddhist-Madhyamaka "philosophy" with its reductionist (logical) negativism.

References

Bstan-Dzin-Rgya-Mtsho; Hayward, J.; Varela, F. (2001). *Gentle bridges : Conversations with the Dalai Lama on the sciences of mind.* Shambhala.

Dalai Lama; Engel, J.; Gackenback; & Varela F. (1997). *Dreaming, sleeping, and dying: An exploration of consciousness with the Dalai Lama.* Wisdom Books.

Guenther, H.V. (1989). *From reductionism to creativity: rDzogs-chen and the new sciences of mind.* Boston: Shambhala.

Maturana, H., & Varela, F. (1980). *Autopoiesis and cognition.* Boston: D. Reidel.

Maturana, H., & Varela, F. (1987). *The tree of knowledge.* Boston: New Science Library.

Varela, F., Maturana, H. R. and R. Uribe (1974). Autopoiesis: The organization of living systems, its characterization and a model. *Biosystems, Vol. 5, No. 4.*

Self

Were mind and matter me,
I would come and go like them.
If I were something else,
They would say nothing about me.

What is mine
When there is no me?
Were self centeredness eased,
I would not think of me and mine –
There would be no one there
To think them.

What is inside is me,
What is outside is mine –
When these thoughts end,
Compulsion stops,
Repetition ceases,
Freedom dawns.

Fixations spawn thoughts
That provoke compulsive acts –
Emptiness stops fixations.

Buddhas speak of "self"
And also teach "no self"
And also say "there's nothing
Which is either self or not."

When things dissolve,
There's nothing left to say.
The unborn and unceasing
Are already free.

Buddha said: "It is real,"
And "it is unreal,"
And "it is both real and unreal,"
And "it is neither one nor the other."

It is all at ease,
Unfixatable by fixations,
Incommunicable,
Inconceivable,
Indivisible.

You are not the same as or different from
Conditions on which you depend;
You are neither severed from
Nor forever fused with them –

This is the deathless teaching
Of buddhas who care for the world.

When buddhas don't appear
And their followers are gone,
The wisdom of awakening
Bursts forth by itself.

Nagarjuna[1]

[1] Trans. Stephen Batchelor, from Nagarjuna, *Verses from the Center: A Buddhist Vision of the Sublime* (NY: Riverhead Books)

Cybernetics & Human Knowing, Vol.9, No.2, 2002, pp. 95–96

Afterword

Klaus Krippendorff[1]

Francisco J. Varela was a student and collaborator of Humberto R. Maturana. Their pioneering collaboration on *Autopoiesis and Cognition* reestablished "processes of living" as the principle topic of biological explorations. This topic had dropped out of the discourse of biology after the work of Jacob von Uexküll. Autopoiesis brought a new framework to biology. I say framework because it was not a theory that predicted observable phenomena but a scaffold to pose and answer new kinds questions. In their *The Tree of Knowledge*, which connected the notion of autopoiesis to a variety of biological, evolutionary, cognitive, and, in a rudimentary way, linguistic and social phenomena, Francisco started to identify his contributions.

Francisco became well known for his *A Calculus for Self-Reference*, which took off from Spencer-Brown's *Laws of Form*. This calculus did not produce a major breakthrough in biology, however, probably due to its mathematical nature, which was too far removed from the messy complexity of biological phenomena. But the ideas that underlie both autopoiesis and self-reference gave rise to Francisco's *Principles of Biological Autonomy* in which he outlines a research program for biology that challenges many cherished concepts, perhaps too many to be widely embraced, perhaps too far ahead of its time. He saw autopoiesis as one manifestation of autonomy and looked for and found autonomous systems in many living systems.

For example, he investigated the pathways of visual perception in mammals, commonly theorized as sequentially transmitting information from the retina to the brain. He found that the lateral geniculate nucleus (LGN), thought to be some kind of converter of visual information, in fact receives only 20% of its information from the eye and 80% from inside the body, including from the brain. It would follow that the brain "sees" mainly its own activity, perturbed by what happens on the retina. For Francisco this finding puts the information processing model of cognition in serious doubt, questions the idea of a semiotic based on the signifier/signified distinction, and challenges the widespread belief that we could be instructed or informed by phenomena outside of us. He therefore abandoned the use of information as an explanatory concept in favor of a notion of in-formation, the transformation that an autonomous system undergoes on its own account.

[1] Gregory Bateson Term Professor for Cybernetics, Language, and Culture, The Annenberg School for Communication, University of Pennsylvania, Philadelphia

Francisco was led to the above from recognizing the connection between autonomy as a fundamental property of living systems and the circular organization necessary to maintain autonomy. This brought Francisco through cybernetics to cognitive science where he added another dimension to this discourse: embodiment, which is alien to mathematics, ignored by the computational theories of Artificial Intelligence, and invisible to the scientific observer as a Cartesian spectator or renaissance kind. Undoubtedly, taking this turn had much to do with his increasing interest in Buddhist meditation and its emphasis on practice rather than abstract theory.

He wrote a small book on *Cognitive Science; A Cartography of Current Ideas*, which gives a history of Artificial Intelligence, offers a decisive critique of the symbol manipulation model of cognition, and proposes emergence as a viable alternative. Finally, it weaves emergence, self-organization, and ontogenesis — the process of constructing reality — into a proposal for a new kind of cognitive science. Subsequently, he co-authored *The Embodied Mind* with Evan Thompson and Eleanor Rosch, extending his proposal to overcome representationalism, information processing, and symbol manipulation conceptions of the mind, bringing cognition closer to experiences.

I could mention several other paths that Francisco took, his reflections on know-how, for example, and on ethical knowledge, much of it must now be read as an invitation to others for to continue, but I will stop here.

Francisco started with the purest form of theory, with a mathematical calculus, and ended in a Herculean effort to recover the human body that abstract theory ignores. Along this path he provided us with numerous revolutionary concepts that changed our conversations. This path is not too different from Ludwig Wittgenstein's who started with a logical Tractatus purporting to solve all philosophical problems with what we now call a picture theory of language and ended in a concept of language as a multitude of games that we invent among ourselves as we go on in life, implicating our body at each turn.

Sadly, Francisco left us far too early, but this was the path he brought forth in walking. I am glad mine joined his on several occasions. I find his work inspiring even in different spaces.

Journal of Consciousness Studies

imprint-academic.com/jcs

'With *JCS*, consciousness studies has arrived'
Susan Greenfield, *Times Higher Education Supplement*

CYBERNETICS
& HUMAN KNOWING

a journal of second-order cybernetics, autopoiesis and cyber-semiotics

CYBERNETICS
& HUMAN KNOWING

a journal of second-order cybernetics,
autopoiesis and cyber-semiotics

10, No.3-4,

CYBERNETICS
& HUMAN KNOWING

a journal of second-order cybernetics,
autopoiesis and cyber-semiotics

Volume 13, No. 1, 2006

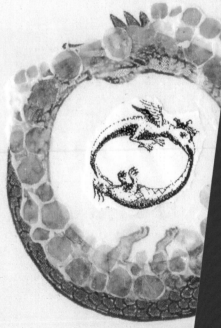

Heinz von Foerst

semiosis & causality

imprint-academic.com/C&HK